BRAINOLOGY

The Curious Science of Our Minds

Contents

Ouch! The science of pain

■ John Walsh

One night in May, my wife sat up in bed and said, 'I've got this awful pain just here.' She prodded her abdomen and made a face. 'It feels like something's really wrong.' Woozily noting that it was 2am, I asked what kind of pain it was. 'Like something's biting into me and won't stop,' she said.

'Hold on,' I said blearily, 'help is at hand.' I brought her a couple of ibuprofen with some water, which she downed, clutching my hand and waiting for the ache to subside.

An hour later, she was sitting up in bed again, in real distress. 'It's worse now,' she said, 'really nasty. Can you phone the doctor?' Miraculously, the family doctor answered the phone at 3am, listened to her recital of symptoms and concluded, 'It might be your appendix. Have you had yours taken out?' No, she hadn't. 'It could be appendicitis,' he surmised, 'but if it was dangerous you'd be in much worse pain than you're in. Go to the hospital in the morning, but for now, take some paracetamol and try to sleep.'

Barely half an hour later, the balloon went up. She was awakened for the third time, but now with a pain so savage and uncontainable it made her howl like a tortured witch face down on a bonfire. The time for murmured assurances and spousal procrastination was over. I rang a local minicab, struggled into my clothes, bundled her into a dressing gown, and we sped to St Mary's Paddington at just before 4am.

The flurry of action made the pain subside, if only through distraction, and we sat for hours while doctors brought forms to be filled, took her blood pressure and ran tests. A registrar poked a needle into my wife's wrist and said, 'Does that hurt? Does that? How about that?' before concluding: 'Impressive. You have a very high pain threshold.'

The pain was from pancreatitis, brought on by rogue gallstones that had escaped from her gall bladder and made their way, like fleeing convicts, to a refuge in her pancreas, causing agony. She was given a course of antibiotics and, a month later, had an operation to remove her gall bladder.

'It's keyhole surgery,' said the surgeon breezily, 'so you'll be back to normal very soon. Some people feel well enough to take the bus home after the operation.' His optimism was misplaced. My lovely wife, she of the admirably high pain threshold, had to stay overnight, and came home the following day filled with painkillers; when they wore off, she writhed with suffering. After three days she rang the specialist, only to be told: 'It's not the operation that's causing discomfort – it's the air that was pumped inside you to separate the organs before surgery.' Like all too many surgeons, they had lost interest in the fallout once the operation had proved a success.

During that period of convalescence, as I watched her grimace and clench her teeth and let slip little cries of anguish until a long regimen of combined ibuprofen and codeine finally conquered the pain, several questions came into my head. Chief among them was: Can anyone in the medical profession talk about pain with any authority? From the family doctor to the surgeon, their remarks and suggestions seemed tentative, generalised, unknowing – and potentially dangerous: Was it right for the doctor to tell my wife that her level of pain didn't sound like appendicitis when the doctor didn't know whether she had a high or low pain threshold? Should he have advised her to stay in bed and risk her appendix exploding into peritonitis? How could surgeons predict that patients would feel only 'discomfort' after such an operation when she felt agony – an agony that was aggravated by fear that the operation had been a failure?

I also wondered if there were any agreed words that would help a doctor understand the pain felt by a patient. I thought of my father, a GP in the 1960s with an NHS practice in south London, who used to marvel at the colourful pain symptoms he heard: 'It's like I've been attacked with a stapler'; 'like having rabbits running up and down my spine'; 'it's like someone's opened a cocktail umbrella in my penis...' Few of them, he told me, corresponded to the symptoms listed in a medical textbook. So how should he proceed? By guesswork and aspirin?

There seemed to be a chasm of understanding in human discussions of pain. I wanted to find out how the medical profession apprehends pain – the language it uses for something that's invisible to the naked eye, that can't be measured except by asking for the sufferer's subjective description, and that can be treated only by the use of opium derivatives that go back to the Middle Ages.

§ § §

When investigating pain, the basic procedure for clinics everywhere is to give a patient the McGill Pain Questionnaire. This was developed in the 1970s by two scientists, Dr Ronald Melzack and Dr Warren Torgerson, both of McGill University in Montreal, and is still the main tool for measuring pain in clinics worldwide.

Melzack and his colleague Dr Patrick Wall of St Thomas' Hospital in London had already galvanised the field of pain research in 1965 with their seminal 'gate control theory', a ground-breaking explanation of how psychology can affect the body's perception of pain. In 1984 the pair went on to write Wall and Melzack's *Textbook of Pain*, the most comprehensive reference work in pain medicine. It's gone through five editions and is currently over 1,000 pages long.

In the early 1970s, Melzack began to list the words patients used to describe their pain and classified them into three categories: sensory (which included heat, pressure, 'throbbing' or 'pounding' sensations), affective (which related to emotional effects, such as 'tiring', 'sickening', 'gruelling' or 'frightful') and lastly evaluative (evocative of an experience – from 'annoying' and 'troublesome' to 'horrible', 'unbearable' and 'excruciating').

You don't have to be a linguistic genius to see there are shortcomings in this lexical smorgasbord. For one thing, some words in the affective and evaluative categories seem interchangeable – there's no difference between 'frightful' in the former and 'horrible' in the latter, or between 'tiring' and 'annoying' – and all the words share an unfortunate quality of sounding like a duchess complaining about a ball that didn't meet her standards.

But Melzack's grid of suffering formed the basis of what became the McGill Pain Questionnaire. The patient listens as a list of 'pain descriptors' is read out and has to say whether each word describes their pain – and, if so, to rate the intensity of the feeling. The clinicians then look at the questionnaire and put check marks in the appropriate places. This gives them a number, or a percentage figure, to work with in assessing, later, whether a treatment has brought the patient's pain down (or up).

A more recent variant is the National Initiative on Pain Control's Pain Quality Assessment Scale (PQAS), in which patients are asked to indicate, on a scale of 1 to 10, how 'intense' – or 'sharp', 'hot', 'dull', 'cold', 'sensitive', 'tender', 'itchy', etc – their pain has been over the past week.

The trouble with this approach is the imprecision of that scale of 1 to 10, where a 10 would be 'the most intense pain sensation imaginable'. How does a patient 'imagine' the worst pain ever and give their own pain a number? Middle-class British men who have never been in a war zone may find it hard to imagine anything more agonising than toothache or a tennis injury. Women who have experienced childbirth may, after that experience, rate everything else as a mild 3 or 4.

I asked some friends what they thought the worst physical pain might be. Inevitably, they just described nasty things that had happened to them. One man nominated gout. He recalled lying on a sofa, with his gouty foot resting on a pillow, when a visiting aunt passed by; the chiffon scarf she was wearing slipped from her neck and lightly touched his foot. It was 'unbearable agony'. A brother-in-law nominated post-root canal toothache – unlike muscular or back pain, he said, it couldn't be alleviated by shifting your posture. It was 'relentless'. A male friend confided that a haemorrhoidectomy had left him with irritable bowel syndrome, in which a daily spasm made him feel 'as if somebody had shoved a stirrup pump up my arse and was pumping furiously'. The pain was, he said, 'boundless, as if it wouldn't stop until I exploded'. A woman friend recalled the moment the hem of her husband's trouser leg snagged on her big toe, ripping the nail clean off. She used a musical analogy to explain the effect: 'I'd been through childbirth, I'd broken my leg – and I recalled them both as low moaning noises, like cellos; the ripped-off nail was excruciating, a great, high, deafening shriek of psychopathic violins, like nothing I'd heard – or felt – before.'

A novelist friend who specialises in World War I drew my attention to Stuart Cloete's memoir *A Victorian Son* (1972), in which the author records his time in a field hospital. He marvels at the stoicism of the wounded soldiers: 'I have heard boys on their stretchers crying with weakness, but all they ever asked for was water or a cigarette. The exception was a man hit through the palm of the hand. This I believe to be the most painful wound there is, as the sinews of the arm contract, tearing as if on a rack.'

Is it true? Looking at the Crucifixion scene in Matthias Grünewald's Isenheim Altarpiece (1512–16), you take in the horribly straining fingers of Christ, twisted around the fat nailheads that skewer his hands to the wood – and oh, God yes, you believe it must be true.

It seems a shame that these eloquent descriptions are reduced by the McGill Questionnaire to words like 'throbbing' or 'sharp', but its function is simply to give pain a number – a number that will, with luck, be decreased after treatment, when the patient is reassessed.

This procedure doesn't impress Professor Stephen McMahon of the London Pain Consortium, an organisation formed in

2002 to promote internationally competitive research into pain. 'There are lots of problems that come with trying to measure pain,' he says. 'I think the obsession with numbers is an over-simplification. Pain is not unidimensional. It doesn't just come with scale – a lot or a little – it comes with other baggage: how threatening it is, how emotionally disturbing, how it affects your ability to concentrate. The measuring obsession probably comes from the regulators who think that, to understand drugs, you have to show efficacy. And the American Food and Drug Administration don't like quality-of-life assessments; they like hard numbers. So we're thrown back on giving it a number and scoring it. It's a bit of a wasted exercise because it's only one dimension of pain that we're capturing.'

§ § §

Pain can be either acute or chronic, and the words do not (as some people think) mean 'bad' and 'very bad'. 'Acute' pain means a temporary or one-off feeling of discomfort, which is usually treated with drugs; 'chronic' pain persists over time and has to be lived with as a malevolent everyday companion. But because patients build up a resistance to drugs, other forms of treatment must be found for it.

The Pain Management and Neuromodulation Centre at Guy's and St Thomas' Hospital in central London is the biggest pain centre in Europe. Heading the team there is Dr Adnan Al-Kaisy, who studied medicine at the University of Basrah, Iraq, and later worked in anaesthetics at specialist centres in England, the USA and Canada.

Who are his patients and what kind of pain are they generally suffering from? 'I'd say that 55 to 60 per cent of our patients suffer from lower back pain,' he says. 'The reason is, simply, that we don't pay attention to the demands life makes on us, the way we sit, stand, walk and so on. We sit for hours in front of a computer, with the body putting heavy pressure on small joints in the back.' Al-Kaisy reckons that in the UK the incidence of chronic lower back pain has increased substantially in the last 15–20 years, and that 'the cost in lost working days is about £6–7 billion'.

Elsewhere the clinic treats those suffering from severe chronic headaches and injuries from accidents that affect the nervous system.

Do they still use the McGill Questionnaire? 'Unfortunately yes,' says Al-Kaisy. 'It's a subjective measurement. But pain can be magnified by a domestic argument or trouble at work, so we try to find out about the patient's life – their sleeping patterns, their ability to walk and stand, their appetite. It's not just the patient's condition, it's also their environment.'

The challenge is to transform this information into scientific data. 'We're working with Professor Raymond Lee, Chair of Biomechanics at the South Bank University, to see if there can be objective measurement of a patient's disability due to pain,' he says. 'They're trying to develop a tool, rather like an accelerometer, which will give an accurate impression of how active or disabled they are, and tell us the cause of their pain from the way they sit or stand. We're really keen to get away from just asking the patient how bad their pain is.'

Some patients arrive with pains that are far worse than backache and require special treatment. Al-Kaisy describes one patient – let us call him Carter – who suffered from a terrible condition called ilioinguinal neuralgia, a disorder that produces a severe burning and stabbing pain in the groin. 'He'd had an operation in the testicular area, and the inguinal nerve had been cut. The pain was excruciating: when he came to us, he was on four or five different medications, opiates with very high dosages, anticonvulsive medication, opioid patches, paracetamol and ibuprofen on top of that. His life was turned upside down, his job was on the line.' The utterly stricken Carter was to become one of Al-Kaisy's big successes.

Since 2010, Guy's and St Thomas' has offered a residential programme for adults whose chronic pain hasn't responded to treatment at other clinics. The patients come in for four weeks, away from their normal environment, and are seen by a motley crew of psychologists, physiotherapists, occupational health specialists and nursing physicians who between them devise a programme to teach them strategies for managing their pain.

Many of these strategies come under the heading of 'neuro-modulation', a term you hear everywhere in pain management circles. In simple terms, it means distracting the brain from constantly brooding on the pain signals it's getting from the body's 'periphery'. Sometimes the distraction is a cunningly deployed electric shock.

'We were the first centre in the world to pioneer spinal cord stimulation,' says Al-Kaisy proudly. 'In pain occasions, over-active nerves send impulses from the periphery to the spinal cord and from there to the brain, which starts to register pain. We try to send small bolts of electricity to the spinal cord by inserting a wire in the epidural area. It's only one or two volts, so the patient feels just a tingling sensation over where the pain is, instead of feeling the actual pain. After two weeks, we give the patient an internal power battery with a remote control, so he can switch it on whenever he feels pain and carry on with his life. It's essentially a pacemaker that suppresses the hyperexcitability of nerves by delivering subthreshold stimulation. The patient feels nothing except his pain going down. It's not invasive – we usually send patients home the same day.'

When Carter, the chap with the agonised groin, had failed to respond to any other treatments, Al-Kaisy tried his box of tricks. 'We gave him something called a dorsal root ganglion stimulation. It's like a small junction-box, placed just underneath one of the bones of the spine. It makes the spine hyperexcited, and sends impulses to the spinal cord and the brain. I pioneered a new technique to put a small wire into the ganglion, connected to an external power battery. Over ten days the intensity of pain went down by 70 per cent – by the patient's own assessment. He wrote me a very nice email saying I had changed his life, that the pain had just stopped completely, and that he was coming back to normality. He said his job was saved, as was his marriage, and he wanted to go back to playing sport. I told him, 'Take it easy. You mustn't start climbing the Himalayas just yet." Al-Kaisy beams. 'This is a remarkable outcome. You cannot get it from any other therapies.'

§ § §

The greatest recent breakthrough in assessing pain, according to Professor Irene Tracey, head of the University of Oxford's Nuffield Department of Clinical Neurosciences, has been the understanding that chronic pain is a thing in its own right. She explains: 'We always thought of it as acute pain that just goes on and on – and if chronic pain is just a continuation of acute pain, let's fix the thing that caused the acute and the chronic should go away. That has spectacularly failed. Now we think of chronic pain as a shift to another place, with different mechanisms, such as changes in genetic expression, chemical release, neurophysiology and wiring. We've got all these completely new ways of thinking about chronic pain. That's the paradigm shift in the pain field.'

Tracey has been called the 'Queen of Pain' by some media commentators. She was, until recently, the Nuffield Professor of Anaesthetic Science and is an expert in neuroimaging techniques that explore the brain's responses to pain. Despite her nickname, in person she is far from alarming: a bright-eyed, enthusiastic, welcoming and hectically fluent woman of 50, she talks about pain at a personal level. She has no problem defining the 'ultimate pain' that scores 10 on the McGill Questionnaire: 'I've been through childbirth three times, and my 10 is a very different 10 from before I had kids. I've got a whole new calibration on that scale.' But how does she explain the ultimate pain to people who haven't experienced childbirth? 'I say, 'Imagine you've slammed your hand in a car door – that's 10.''

She uses a personal example to explain the way perception and circumstance can alter the way we experience pain, as well as the phenomenon of 'hedonic flipping', which can convert pain from an unpleasant sensation into something you don't mind. 'I did the London Marathon this year. It needs a lot of training and running and your muscles ache, and next day you're really in pain, but it's a nice pain. I'm no masochist, but I associate the muscle pain with thoughts like, 'I did something healthy with my body,' 'I'm training,' and 'It's all going well.''

I ask her why there seems to be a gap between doctors' and patients' apprehension of pain. 'It's very hard to understand, because the system goes wrong from the point of injury, along the nerve that's taken the signal into the spinal cord, which sends signals to the brain, which sends signals back, and it all unravels with terrible consequential changes. So my patient may be saying, 'I've got this excruciating pain here,' and I'm trying to see where it's coming from, and there's a mismatch here because you can't see any damage or any oozing blood. So we say, 'Oh come now, you're obviously exaggerating, it can't be as bad as that.' That's wrong – it's a cultural bias we grew up with, without realising.'

Recently, she says, there has been an explosion of understanding about how the brain is involved in pain. Neuroimaging, she explains, helps to connect the subjective pain with the objective perception of it. 'It fills that space between what you can see and what's being reported. We can plug that gap and explain why the patient is in pain even though you can't see it on your X-ray or whatever. You're helping to bring truth and validity to these poor people who are in pain but not believed.'

But you can't simply 'see' pain glowing and throbbing on the screen in front of you. 'Brain imaging has taught us about the networks of the brain and how they work,' she says. 'It's not a pain-measuring device. It's a tool that gives you fantastic insight into the anatomy, the physiology and the neurochemistry of your body and can tell us why you have pain, and where we should go in and try to fix it.'

Some of the ways in, she says, are remarkably direct and mechanical – like Al-Kaisy's spinal cord stimulation wire. 'There are now devices you can attach to your head and allow you to manipulate bits of the brain. You can wear them like bathing caps. They're portable, ethically allowed brain-simulation devices. They're easy for patients to use and evidence is coming, in clinical trials, that they are good for strokes and rehabilitation. There's a parallel with the games industry, where they're making devices you can put on your head so kids can use thought to move balls around. The games industry is, for fun, driving this idea that when you use your brain, you generate electrical activities. They're developing the technology really fast, and we can use it in medical applications.'

§ § §

According to the International Association for the Study of Pain, pain is defined as 'an unpleasant sensory and emotional experience associated with actual or potential tissue damage, or described in terms of such damage'. It's a broad-brush definition that hints at the holistic nature of pain and the range of factors that might influence our perception of it. If not all of its causes are directly physical, standardised drug treatments will always be something of a blunt instrument.

Researchers at the Human Pain Research Laboratory at Stanford University, California, are working to gain a better understanding of individual responses to pain so that treatments can be more targeted. The centre was created in 1995 by the pleasingly named Dr Martin Angst of the Department of Anesthesiology. Its first investigations were into finding reliable methods of quantifying pain. Then Angst (assisted by the equally pleasingly named Dr Martha Tingle) looked into questions of opiate pharmacology, such as how easily the body builds up toleration to drugs.

Pain has become a huge area of medical research in the USA, for a simple reason. Chronic pain affects over 100 million Americans and costs the country over half a trillion dollars a year in lost working hours, which is why it's become a magnet for funding by big business and government.

The laboratory has several study initiatives on the go – into migraine, fibromyalgia, facial pain and other conditions – but its largest is into back pain. It has been endowed with a $10m grant from the National Institutes of Health to study non-drug alternative treatments for lower back pain. The specific treatments are mindfulness, acupuncture, cognitive behavioural therapy and real-time neural feedback. This may seem a very Californian range of pursuits, but the lab takes them very seriously and is enlisting an army of patients to build up a massive database.

They plan to inspect the pain tolerance of 400 people over five years of study, ranging from pain-free volunteers to the most wretched chronic sufferers who have been to other specialists but found no relief. Subjects are all called in, given screening tests (to exclude those with abnormal drug regimens or excessive 'suicidality') then subjected to several quantitative sensory tests: participants are asked to immerse one naked foot in a bucket of iced water until they feel pain; then one arm is subjected to a 'contact heat evoked potential simulator', which gradually heats up small-diameter nerve fibres until the patient feels pain; then they have 'pressure needles' poked onto their skin without breaking it until they report discomfort.

In all three cases, the idea is to find people's mid-range tolerance (they're asked to rate their pain while they're experiencing it), to establish a usable baseline. They then are given the non-invasive treatments – mindfulness, acupuncture, etc – and are subjected afterwards to the same pain stimuli, to see how their pain tolerance has changed from their baseline reading. MRI scanning is used on the patients in both laboratory sessions, so that clinicians can see and draw inferences from the visible differences in blood flow to different parts of the brain.

A remarkable feature of the assessment process is that patients are also given scores for psychological states: a scale measures their level of depression, anxiety, anger, physical functioning, pain behaviour and how much pain interferes with their lives. This should allow physicians to use the information to target specific treatments. All these findings are stored in an 'informatics platform' called CHOIR, which stands for the Collaborative Health Outcomes Information Registry. It has files on 15,000 patients, 54,000 unique clinic visits and 40,000 follow-up meetings.

The big chief at the Human Pain Research Laboratory is Dr Sean Mackey, Redlich Professor of Anesthesiology, Perioperative and Pain Medicine, Neurosciences and Neurology at Stanford. His background is in bioengineering, and under his governance the Stanford Pain Management Center has twice been designated a centre of excellence by the American Pain Society. A tall, genial, easy-going man, he is sometimes approached by legal firms who want him to appear in court to state definitively whether their client is or isn't in chronic pain (and therefore justified in claiming absentee benefit). His response is surprising.

'In 2008, I was asked by a law firm to speak in an industrial injury case in Arizona. This poor guy got hot burning asphalt sprayed on his arm at work; he had a claim of burning neuropathic pain. The plaintiff's side brought in a cognitive scientist, who scanned his brain and said there was conclusive evidence that he had chronic pain. The defence asked me to comment, and I said, 'That's hogwash, we cannot use this technology for that purpose.'

'Shortly afterwards, I gave a talk on pain, neuroimaging and the law, explaining why you can't do this – because there's too much individual variability in pain, and the technology isn't sensor-specific enough. But I concluded by saying, 'If you were to do this, you'd use modern machine-learning approaches, like those used for satellite reconnaissance to determine whether a satellite is seeing a tank or a civilian truck.' Some of my students said, 'Can you give us some money to try this?' I said, 'Yes, but it can't be done.' But they designed the experiment – and discovered that, using brain imagery, they could predict with 80 per cent accuracy whether someone was feeling heat pain or not.'

Mackey finally published a paper about the experiment. So did his findings influence any court decisions? 'No. I get asked by attorneys, and I always say, 'There is no place for this in the courtroom in 2016 and there won't be in 2020. People want to push us into saying this is an objective biomarker for detecting that someone's in pain. But the research is in carefully controlled laboratory conditions. You cannot generalise about the population as a whole. I told the attorneys, 'This is too much of a leap.' I don't think there's a lot of clinical utility in having a pain-o-meter in a court or in most clinical situations.'

Mackey explains the latest thinking about what pain actually is. 'Now we understand that pain is a balance between ascending information coming from our bodies and descending inhibitory systems from our brains. We call the ascending information 'nociception' – from the Latin nocere, to harm or hurt – meaning the response of the sensory nervous system to potentially harmful stimuli coming from our periphery, sending signals to the spinal cord and hitting the brain with the perception of pain. The descending systems are inhibitory, or filtering, neurons, which exist to filter out information that's not important, to 'turn down' the ascending signals of hurt.

The main purpose of pain is to be the great motivator, to tell you to pay attention, to focus. When Martin was doing the pain lab, we had no way of addressing these two dynamic systems, and now we can.'

Mackey is immensely proud of his massive CHOIR database – which records people's pain tolerance levels and how they are affected by treatment – and has made it freely available to other pain clinics as a 'community source platform', collaborating with academic medical centres nationwide 'so that a rising tide elevates all boats'. But he's also humble enough to admit that science can't tell us which are the sites of the body's worst pains.

'Back pain is the most reported pain at 28 per cent, but I know there's a higher density of nerve fibres in the hands, face, genitals and feet than in other areas. And there are conditions where the sufferer has committed suicide to get away from the pain: things like post-herpetic neuralgia, that burning nerve pain that occurs after an outbreak of shingles and is horrific; another is cluster headaches – some patients have thought about taking a drill to their heads to make it stop.'

Like Irene Tracey, he's enthusiastic about the rise of transcranial magnetic stimulation ('Imagine hooking a nine-volt battery across your scalp') but, when asked about his particular successes, he talks about simple solutions. 'Early on in my career, I used to be very focused on the peripheral, the apparent site of the pain. I was doing interventions, and some people would get better but a lot wouldn't. So I started listening to their fears and anxieties and working on those, and became very brain-focused. I noticed that if you have a nerve trapped in your knee, your whole leg could be on fire, but if you apply a local anaesthetic there, it could abolish it.

'This young woman came to me with a terrible burning sensation in her hand. It was always swollen; she couldn't stand anyone touching it because it felt like a blowtorch.' Mackey noticed that she had a post-operative scar from prior surgery for carpal tunnel syndrome. Speculating that this was at the root of her problem, he injected Botox, a muscle relaxant, at the site of the scar. 'A week later, she came up and gave me this huge hug and said, 'I was able to pick up my child for the first time in two years. I haven't been able to since she was born.' All the swelling was gone. It taught me that it's not all about the body part, and not all about the brain. It's about both.' How counterintuitive to discover that, after centuries of curing pain with opiates, the mind can give the morphine a run for its money.

This story was first published on 10 January 2017
by Wellcome on mosaicscience.com

Why doctors are reclaiming LSD and ecstasy

■ Sam Wong

At 6.30am on Thursday 29 October 2009, Friederike Meckel Fischer's doorbell rang. There were ten policemen outside. They searched the house, put handcuffs on Friederike – a diminutive woman in her 60s – and her husband, and took them to a remand prison. The couple had their photographs and fingerprints taken and were put in separate cells in isolation. After a few hours, Friederike, a psychotherapist, was taken for questioning.

The officer read back to her the promise of secrecy she had each client make at the start of her group therapy sessions. 'Then I knew I was really in trouble,' she says.

'I promise not to divulge the location or names of the people present or the medication. I promise not to harm myself or others in any way during or after this experience. I promise that I will come out of this experience healthier and wiser. I take personal responsibility for what I do here.'

The Swiss police had been tipped off by a former client whose husband had left her after they had attended therapy. She held Friederike responsible.

What got Friederike in trouble were her unorthodox therapy methods. Alongside separate sessions of conventional talk therapy, she offered a catalyst, a tool to help her clients reconnect with their feelings, with people around them, and with difficult experiences in their lives. That catalyst was LSD. In many of her sessions, they would also use another substance: MDMA, or ecstasy.

Friederike was accused of putting her clients in danger, dealing drugs for profit, and endangering society with 'intrinsically dangerous drugs'. Such psychedelic therapy is on the fringes of both psychiatry and society. Yet LSD and MDMA began life as medicines for therapy, and new trials are testing whether they could be again.

§§§

In 1943, Albert Hofmann, a chemist at the Sandoz pharmaceutical laboratory in Basel, Switzerland, was trying to develop drugs to constrict blood vessels when he accidentally ingested a small quantity of lysergic acid diethylamide, LSD. The effects shook him. As he writes in his book LSD, *My Problem Child*:

'Objects as well as the shape of my associates in the laboratory appeared to undergo optical changes... Light was so intense as to be unpleasant. I drew the curtains and immediately fell into a peculiar state of 'drunkenness', characterised by an exaggerated imagination. With my eyes closed, fantastic pictures of extraordinary plasticity and intensive colour seemed to surge towards me. After two hours, this state gradually subsided and I was able to eat dinner with a good appetite.'

Intrigued, he decided to take the drug a second time in the presence of colleagues, an experiment to determine whether it was indeed the cause. The faces of his colleagues soon appeared 'like grotesque coloured masks', he writes:

'I lost all control of time: space and time became more and more disorganised and I was overcome with fears that I was going crazy. The worst part of it was that I was clearly aware of my condition though I was incapable of stopping it. Occasionally I felt as being outside my body. I thought I had died. My 'ego' was suspended somewhere in space and I saw my body lying dead on the sofa. I observed and registered clearly that my 'alter ego' was moving around the room, moaning.'

But he seemed particularly struck by what he felt the next morning: 'Breakfast tasted delicious and was an extraordinary pleasure. When I later walked out into the garden, in which the sun shone now after a spring rain, everything glistened and sparkled in a fresh light. The world was as if newly created. All my senses vibrated in a condition of highest sensitivity that persisted for the entire day.'

Hofmann felt it was of great significance that he could remember the experience in detail. He believed the drug could hold tremendous value to psychiatry. The Sandoz labs, after ensuring it was non-toxic to rats, mice and humans, soon started offering it for scientific and medical use.

One of the first to start using the drug was Ronald Sandison. The British psychiatrist visited Sandoz in 1952 and, impressed by Hofmann's research, left with 100 vials of what was by then called Delysid. Sandison immediately began giving it to patients at Powick Hospital in Worcestershire who were failing to make progress in traditional psychotherapy. After three years, the hospital bosses were so pleased with the results that they built a new LSD clinic. Patients would arrive in the morning, take their LSD, then lie down in private rooms. Each had a record player and a blackboard for drawing on, and nurses or registrars would check on them regularly. At 4pm the patients would convene and discuss their experiences, then a driver would take them home, sometimes while they were still under the influence of the drug.

Around the same time, another British psychiatrist, Humphry Osmond, working in Canada, experimented with using LSD to help alcoholics stop drinking. He reported that the drug, in combination with supportive psychiatry, achieved abstinence rates of 40–45 per cent – far higher than any other treatment at the time or since. Elsewhere, studies of people with terminal cancer showed that LSD therapy could relieve severe pain, improve quality of life and alleviate the fear of death.

In the USA, the CIA tried giving LSD to unsuspecting members of the public to see if it would make them give up secrets. Meanwhile at Harvard University, Timothy Leary – encouraged by, among others, the beat poet Allen Ginsberg – gave it to artists and writers, who would then describe their experiences. When rumours spread that he was giving drugs to students, law-enforcement officials started investigating and the university warned students against taking the drug. Leary took the opportunity to preach about the drug's power as an aid to spiritual development, and was soon sacked from Harvard, which further fuelled his and the drug's notoriety. The scandal had caught the eye of the press and soon the whole country had heard of LSD.

By 1962, Sandoz was cutting back on its distribution of LSD, the result of restrictions on experimental drug use brought on by an altogether different drug scandal: birth defects linked to the morning-sickness drug thalidomide. Paradoxically, the restrictions coincided with an increase in LSD's availability – the formula was not difficult or expensive to obtain, and those who were determined to could synthesise it with moderate difficulty and in great amounts.

Still, moral panic about its effects on young minds was rife. The authorities were also worried about LSD's association with the counterculture movement and the spread of anti-authoritarian views. Calls for a nationwide ban soon followed, and many psychiatrists stopped using LSD as its negative reputation grew.

One of many stories in the press told of Stephen Kessler, who murdered his mother-in-law and claimed afterwards that he didn't remember what he'd done as he was 'flying on LSD'. In the trial, it emerged that he had taken LSD a month earlier, and at the time of the murder was intoxicated only with alcohol and sleeping pills, but millions believed that LSD had turned him into a killer. Another report told of college students who went blind after staring at the sun on LSD.

Two US Senate subcommittees held in 1966 heard from doctors who claimed that LSD caused psychosis and 'the loss of all cultural values', as well as from LSD supporters such as Leary and Senator Robert Kennedy, whose wife Ethel was said to have undergone LSD therapy. 'Perhaps to some extent we have lost sight of the fact that it can be very, very helpful in our society if used properly,' said Kennedy, challenging the Food and Drug Administration for shutting down LSD research programmes.

Possession of LSD was made illegal in the UK in 1966 and in the USA in 1968. Experimental use by researchers was still possible with licences, but with the stigma attached to the drug's legal status, these became extremely hard to get. Research ground to a halt, but illegal recreational use carried on.

§§§

At the age of 40, after 21 years of marriage, Friederike Meckel Fischer fell in love with another man. Sadly, as she soon discovered, he was using her to get out of his own marriage. 'I had a pain within myself with this man having left me, with my husband whom I couldn't connect to,' she says. 'It was just like I was out of myself.'

Her solution was to become a psychotherapist. She says she never thought of going into therapy herself, which in 1980s West Germany was reserved for only the most serious conditions. Besides which, her upbringing taught her to do things herself rather than seek help from others.

Friederike was at the time working as an occupational physician. She recognised that many of the problems she saw in her patients were rooted in problems with their bosses, colleagues or families. 'I came to the conclusion that everything they were having trouble with was connected to relationship issues,' she says.

A former professor of hers recommended she try a technique called holotropic breathwork. Developed by Stanislav Grof, one of the pioneers of LSD psychotherapy, this is a way to induce altered states of consciousness through accelerated and deeper breathing, like hyperventilation. Grof had developed holotropic breathwork in response to bans on LSD use around the world.

Over three years, travelling back and forth to the USA on holidays, Friederike underwent training with Grof as a holotropic breathwork facilitator. At the end of it, Grof encouraged her to try psychedelics.

In the last seminar, a colleague gave her two little blue pills as a gift. When she got back to Germany, Friederike shared one of the blue pills with her friend Konrad, who later became her husband. She says she felt herself lifted by a wave and thrown onto a white beach, able to access parts of her psyche that were off-limits before. 'The first experience was breathtaking for me,' she says. 'I only thought: 'That's it. I can see things.' And I started feeling. That was, for me, unbelievable.'

The pills were MDMA, a drug which had entered the spotlight in 1976 when American chemist Alexander 'Sasha' Shulgin rediscovered it 62 years after it was patented by Merck and then forgotten. In a story echoing that of LSD's origins, upon taking it, Shulgin noted feelings of 'pure euphoria' and 'solid inner strength', and felt he could 'talk about deep or personal subjects with special clarity'. He introduced it to his friend Leo Zeff, a retired psychotherapist who had worked with LSD and believed the obligation to help patients took priority over the law. Zeff had continued to work with LSD secretly after its prohibition. MDMA's potential brought Zeff out of retirement. He travelled around the USA and Europe to instruct therapists on MDMA therapy. He called it 'Adam' because it put the patient in a primordial state of innocence, but at the same time, it had acquired another name in nightclubs: ecstasy.

MDMA was made illegal in the UK by a 1977 ruling that put the entire chemical family in the most tightly controlled category: class A. In the USA, the Drug Enforcement Administration (DEA), set up by Richard Nixon in 1973, declared a temporary ban in 1985. At a hearing to decide its permanent status, the judge recommended that it should be placed in schedule three, which would allow use by therapists. But the DEA overruled the judge's decision and put MDMA in schedule one, the most restrictive category. Under American influence, the UN Commission on Narcotic Drugs gave MDMA a similar classification under international law (though an expert committee formed by the World Health Organization argued that such severe restrictions were not warranted).

Schedule one substances are permitted to be used in research under the UN Convention on Psychotropic Substances. In Britain and the USA, researchers and their institutions must apply for special licences, but these are expensive to obtain, and finding manufacturers who will supply controlled drugs is difficult.

But in Switzerland, which at the time was not a signatory to the convention, a small group of psychiatrists persuaded the government to permit the use of LSD and MDMA in therapy. From 1985 until the mid-1990s, licensed therapists were permitted to give the drugs to any patients, to train other therapists in using the drugs, and to take them themselves, with little oversight.

Believing that MDMA might help her gain a deeper understanding of her own problems, Friederike applied for a place on a 'psycholytic therapy' course in Switzerland. In 1992, she and Konrad were accepted into a training group run by a licensed therapist named Samuel Widmer.

The course took place on weekends every three months at Widmer's house in Solothurn, a town west of Zurich. Central to the training was taking the substances a number of times, 12 altogether, to get to know their effects and go through a process of self-exploration. Friederike says the drug experiences showed her how her whole life had been coloured by the loss of her father at the age of 5 and the hardship of growing up in postwar West Germany.

'I can detect relations, interconnections between things that I couldn't see before,' she says of her experiences with MDMA. 'I could look at difficult experiences in my life without getting right away thrown into them again. I could for example see a traumatic experience but not connect to the horrible feeling of the moment. I knew it was a horrible thing, and I could feel that I have had fear but I didn't feel the fear.'

§ § §

People on psychedelic highs often speak of profound, spiritual experiences. Back in the 1960s, Walter Pahnke, a student of Timothy Leary, conducted a notorious experiment at Boston University's Marsh Chapel showing that psychedelics could induce these.

He gave ten volunteers a large dose of psilocybin – the active ingredient in magic mushrooms – and ten an active placebo, nicotinic acid, which caused a tingling sensation but no mental effects. Eight of the psilocybin group had spiritual experiences, compared with one of the placebo group. In later studies, researchers have identified core characteristics of such experiences, including ineffability, the inability to put it into words; paradoxicality, the belief that contradictory things are true at the same time; and feeling more connected to other people or things.

'When the experience can be really useful is when they feel a connection even with someone who has caused them hurt, and an understanding of what may have caused them to behave in the way they did,' says Robin Carhart-Harris, a psychedelics researcher at Imperial College London. 'I think the power to achieve those kinds of realisations really speaks to the incredible value of psychedelics and captures why they can be so effective and valuable in therapy. I think that can only really happen when defences dissolve away. Defences get in the way of those realisations.'

He compares the feeling of connection with things beyond oneself to the 'overview effect' felt by astronauts when they look back on the Earth. 'All of a sudden they think, 'How silly of me and people in general to have conflict and silly little hang-ups that we think are massive and important.' When you're up in space looking down on the entirety of the Earth, it puts it into perspective. I think a similar kind of overview is engendered by psychedelics.'

Carhart-Harris is conducting the first clinical trial to study psilocybin as a treatment for depression. He is one of a few researchers across the world who are pushing ahead with research on psychedelic therapy. Twelve people have taken part in his study so far.

They begin with a brain scan, and a long preparation session with the psychiatrists. On the therapy day, they arrive at 9am, complete a questionnaire, and have tests to make sure they haven't taken other drugs. The therapy room has been decorated with drapes, ornaments, coloured glowing lights, electric candles, and an aromatiser. A PhD student, who is also a musician, has prepared a playlist, which the patient can listen to either through headphones or from high-quality speakers in the room. They spend most of the session lying on a bed, exploring their thoughts. Two psychiatrists sit with them, and interact when the patient wants to talk. The patients have two therapy sessions: one with a low dose, then one with a high dose. Afterwards, they have a follow-up session to help them integrate their experiences and cultivate healthier ways of thinking.

I meet Kirk, one of the participants, two months after his high-dose session. Kirk had been depressed, particularly since his mother's death three years ago. He experienced entrenched thought patterns, like going round and round on a racetrack of negative thoughts, he says. 'I wasn't as motivated, I wasn't doing as much, I wasn't exercising any more, I wasn't as social, I was having anxiety quite a bit. It just deteriorated. I got to the point where I felt pretty hopeless. It didn't match really what was going on in my life. I had a lot of good things going on in my life. I'm employed, I've got a job, I've got family, but really it was like a quagmire that you sink into.'

At the peak of the drug experience, Kirk was deeply affected by the music. He surrendered himself to it and felt overcome with awe. When the music was sad, he would think of his mother, who had been ill for many years before her death. 'I used to go to the hospital and see her, and a lot of the time she'd be asleep, so I wouldn't wake her up; I'd just sit on the bed. And she'd be aware I was there and wake up. It was a very loving feeling. Quite intensely I went through that moment. I think that was quite good in a way. I think it helped to let go.'

During the therapy sessions, there were moments of anxiety as the drug's effects started to take hold, when Kirk felt cold and became preoccupied with his breathing. But he was reassured by the therapists, and the discomfort passed. He saw bright colours, 'like being at the funfair', and felt vibrations permeate his body. At one point, he saw the Hindu elephant god Ganesh look in at him, as if checking on a child.

Although the experience had been affecting, he noticed little improvement in his mood in the first ten days afterwards. Then, while out shopping with friends on a Sunday morning, he felt an upheaval. 'I feel like there's space around me. It felt like when my mum was still alive, when I first met my partner, and everything was kind of OK, and it was so noticeable because I hadn't had it in a while.'

There have been ups and downs since, but overall, he feels much more optimistic. 'I haven't got that negativity any more. I'm being more social; I'm doing stuff. That kind of heaviness, that suppressed feeling has gone, which is amazing, really. It's lifted a heavy cloak off me.'

Another participant, Michael, had been battling depression for 30 years, and tried almost every treatment available. Before taking part in the trial, he had practically given up hope. Since the day of his first dose of psilocybin, he has felt completely different. 'I couldn't believe how much it had changed so quickly,' he says. 'My approach to life, my attitude, my way of looking at the world, just everything, within a day.'

One of the most valuable parts of the experience helped him to overcome a deep-rooted fear of death. 'I felt like I was being shown what happens after that, like an afterlife,' he says. 'I'm not a religious person and I'd be hard pushed to say I was anything near spiritual either, but I felt like I'd experienced some of that, and experienced the feeling of an afterlife, like a preview almost, and I felt totally calm, totally relaxed, totally at peace. So that when that time comes for me, I will have no fear of it at all.'

§ § §

During her training with Samuel Widmer, Friederike also worked in an addiction clinic. The insights from her drug experiences gave her new empathy. 'All of a sudden I could understand my clients in the clinic with their alcohol addiction,' she says. 'They were coping differently than I did. They had almost the same problems or symptoms I had, only I hadn't started drinking.' But only a few of them were able to open up about how those experiences made them feel. She wondered: could an MDMA experience help them release those emotions?

MDMA is a tamer relative of the classic psychedelics – psilocybin, LSD, mescaline, DMT. They have effects that can be disturbing, like sensory distortions, the dissolution of one's sense of self, and the vivid reliving of frightening memories. MDMA's effects are shorter-lasting, making it easier to handle in a psychotherapy session.

Friederike opened her own private psychedelic therapy practice in Zurich in 1997. During the next few years, she began hosting weekend group therapy sessions with psychedelics in her home, inviting clients who had failed to make progress in conventional talking therapy.

Since the 1950s, psychiatrists have recognised the importance of context in determining what sort of experience the LSD taker would have. They have emphasised the importance of 'set' – the user's mindset, their beliefs, expectations, and experience – and 'setting' – the physical milieu where the drug is taken, the sounds and features of the environment and the other people present.

A supportive setting and an experienced therapist can lower the risk of a bad trip, but frightening experiences still happen. According to Friederike, they are part of the therapeutic experience. 'If a client is able to go through or lets himself be led through and work through, the bad trip turns into the most important step on the way to himself,' she says. 'But without a correct setting, without a therapist who knows what he's doing and without the commitment of the client, we end up in a bad trip.'

Her clients would come to her house on a Friday evening, talk about their recent issues and discuss what they wanted to achieve in the drug session. On Saturday morning, they would sit in a circle on mats, make the promise of secrecy, and each take a personal dose of MDMA agreed with Friederike in advance. Friederike would start with silence, then play music, and speak to the clients individually or as a group to work through their issues. Sometimes she would ask other members of the group to assume the role of a client's family members, and have them discuss problems in their relationship. In the afternoon they would do the same with LSD, which would often let the participants feel as though they were reliving traumatic memories. Friederike would guide them through the experience, and help them understand it in a new way. On Sunday, they would discuss the experiences of the previous day and how to integrate them into their lives.

Friederike's practice, however, was illegal. Therapeutic licences to use the drugs had been withdrawn by the Swiss government around 1993, following the death of a patient in France under the effect of ibogaine, another psychotropic drug. (It was later determined that she died from an undiagnosed heart condition.)

§ § §

The early LSD researchers had no way to look at what it was doing inside the brain. Now we have brain scans. Robin Carhart-Harris has carried out such studies with psilocybin, LSD and MDMA. He tells me there are two basic principles of how the classic psychedelics work. The first is disintegration: the parts that make up different networks in the brain become less cohesive. The second is desegregation: the systems that specialise for particular functions as the brain develops become, in his words, 'less different' from each other.

These effects go some way to explaining how psychedelics could be therapeutically useful. Certain disorders, such as depression and addiction, are associated with characteristic patterns of brain activity that are difficult to break out of. 'The brain kind of enters these patterns, pathological patterns, and the patterns can become entrenched. The brain easily gravitates into these patterns and gets stuck in them. They are like whirlpools, and the mind gets sucked into these whirlpools and gets stuck.'

Psychedelics dissolve patterns and organisation, introducing 'a kind of chaos', says Carhart-Harris. On the one hand, chaos can be seen as a bad thing, linked with things like psychosis, a kind of 'storm in the mind', as he puts it. But you could also view that chaos as having therapeutic value. 'The storm could come and wash away some of the pathological patterns and entrenched patterns that have formed and underlie the disorder. Psychedelics seem to have the potential through this effect on the brain to dissolve or disintegrate pathologically entrenched patterns of brain activity.'

The therapeutic potential suggested by Carhart-Harris's brain scan studies persuaded the UK's Medical Research Council to fund the psilocybin trial for depression. It's too early to evaluate its success, but the results so far have been encouraging. 'Some patients are in remission now months after having had their treatment,' Carhart-Harris says. 'Previously their depressions were very severe, so I think those cases can be considered transformations. I'm not sure if there are any other treatments out there that really have that potential to transform a patient's situation after just two treatment sessions.'

§ § §

In the wake of MDMA's prohibition, American psychologist Rick Doblin founded the Multidisciplinary Association for Psychedelic Studies (MAPS) to support research aiming to re-establish psychedelics' place in medicine. When Swiss psychiatrist Peter Oehen heard they were funding a study on using MDMA to help people with post-traumatic stress disorder (PTSD), he jumped on a plane to meet Doblin in Boston.

Like Friederike, Oehen trained in psychedelic therapy while it was legal in Switzerland in the early 1990s. Doblin agreed to support a small study with 12 patients at Oehen's private practice in Biberist, a small town about half an hour by train from the Swiss capital, Bern.

Oehen thinks that MDMA's mood-elevating, fear-reducing and pro-social effects make it a promising tool to facilitate psychotherapy for PTSD. 'Many of these traumatised people have been traumatised by some kind of interpersonal violence and have lost their ability to connect, are distrustful, are aloof,' says Oehen. 'This helps them regain trust. It helps build a sound and trustful therapeutic relationship.' It also puts the patient in a state of mind where they can face their traumatic memories without becoming distressed, he says, helping to start reprocess the trauma in a different way.

When MAPS's first PTSD study in the USA was published in 2011, the results were eye-opening. After two psychotherapy sessions with MDMA, 10 out of 12 participants no longer met the criteria for PTSD. The benefits were still apparent when the patients were followed up three to four years after the therapy.

Oehen's results were less dramatic, but all of the patients who had MDMA-assisted therapy felt some improvement. 'I'm still in touch with almost half of the people,' he says. 'I can see still people getting better after years going on in the process and re-solving their problems. We saw this at long-term follow-up, that symptoms get better after time, because the experiences enable them to get better in a different way to normal psychotherapy. These effects – being more open, being more calm, more willing to face difficult issues – this goes on.'

In people with PTSD, the amygdala, a primitive part of the brain that orchestrates fear responses, is overactive. The pre-frontal cortex, a more sophisticated part of the brain that allows rational thoughts to override fear, is underactive. Brain-imaging studies with healthy volunteers have shown that MDMA has the opposite effects – boosting the prefrontal cortex response and shrinking the amygdala response.

Ben Sessa, a psychiatrist working around Bristol in the UK, is preparing to carry out a study at Cardiff University testing whether people with PTSD respond to MDMA in the same way. He believes that early negative experiences lie at the root not just of PTSD but of many other psychiatric disorders too, and that psychedelics give patients the ability to reprocess those memories.

'I've been doing psychiatry for almost 20 years now and every single one of my patients has a history of trauma,' he says. 'Mal-treatment of children is the cause of mental illness, in my opin-ion. Once a person's personality has been formed in childhood and adolescence and into early adulthood, it's very difficult to encourage a patient to think otherwise.' What psychedelics do, more than any other treatment, he says, is offer an opportunity to 'press the reset button' and give the patient a new experience of a personal narrative.

Sessa is planning a separate study to test MDMA as a treat-ment for alcohol dependency syndrome – picking up the trail of

Humphrey Osmond's LSD research 60 years ago. He believes psychiatry would look very different today if research with psychedelics had proceeded unencumbered since the 1950s. Psychiatrists have since turned to antidepressants, mood stabilisers and antipsychotics. These drugs, he says, help to manage a patient's condition, but aren't curative, and also carry dangerous side-effects.

'We've become so used to psychiatry being a palliative care field of medicine,' Sessa says. 'That we're with you for life. You come to us in your early 20s with severe anxiety disorder; I'll still be looking after you in your 70s. We've become used to that. And I think we're selling our patients short.'

Will psychedelic drugs ever be ruled legal medicines again? MAPS are supporting trials of MDMA-assisted psychotherapy for PTSD in the USA, Australia, Canada and Israel, and they hope they will have enough evidence to convince regulators to approve it by 2021. Meanwhile, trials using psilocybin to treat anxiety in people with cancer have been taking place at Johns Hopkins University and New York University since 2007.

Few psychiatrists I asked about the legal use of psychedelics in therapy would give their opinions. One of the few who did, Falk Kiefer, Medical Director at the Department of Addictive Behaviour and Addiction Medicine at the Central Institute of Mental Health in Mannheim, Germany, says he is sceptical about the drugs' ability to change patients' behaviour. 'Psychedelic treatment might result in gaining new insights, 'seeing the world in a different way'. That's fine, but if it does not result in learning new strategies to deal with your real world, the clinical outcome will be limited.'

Carhart-Harris says the only way to change people's minds is for the science to be so good that funders and regulators can't ignore it. 'The idea is that we can present data that really becomes irrefutable, so that those authorities that have reservations, we can start changing their perspective and bring them around to taking this seriously.'

§§§

After 13 days under arrest, Friederike was released. She appeared in court in July 2010, accused of violating the narcotics law and endangering her clients, the latter of which could mean up to 20 years' imprisonment. A number of neuroscientists and psychotherapists testified in her defence, arguing that one portion of LSD is not a dangerous substance and has no significant harmful effects when taken in a controlled setting (MDMA was not included in the prosecution's case).

The judge accepted that Friederike had given her clients drugs as part of a therapeutic framework, with careful consideration for their health and welfare, and ruled her guilty of handing out LSD but not guilty of endangering people. For the narcotics offence, she was fined 2,000 Swiss francs and given a 16-month suspended sentence with two years' probation.

'I have been blessed by a very understanding lawyer and an intelligent judge,' she says. She even considers the woman who reported her to the police a blessing, since the case has allowed her to talk openly about her work with psychedelics. She gives occasional lectures at psychedelic conferences, and has written a book about her experience, which she hopes will guide other therapists in how to work with the substances safely.

This story was first published on 12 January 2016
by Wellcome on mosaicscience.com

Inside the mind of an interpreter

■ Geoff Watts

One morning this summer I paid a visit to the sole United Nations agency in London. The headquarters of the International Maritime Organization (IMO) sit on the southern bank of the Thames, a short distance upstream from the Houses of Parliament. As I approached, I saw that a ship's prow, sculpted in metal, was grafted like a nose to the ground floor of this otherwise bland building. Inside I met a dozen or so mostly female IMO translators. They were cheerful and chatty and better dressed than you might imagine for people who are often heard but rarely seen.

I walked upstairs to a glass-fronted booth, where I prepared to witness something both absolutely remarkable and utterly routine. The booth was about the size of a garden shed, and well lit but stuffy. Below us were the gently curving desks of the delegate hall, which was about half-full, occupied mostly by men in suits. I sat down between two interpreters named

Marisa Pinkney and Carmen Soliño, and soon the first delegate started talking. Pinkney switched on her microphone. She paused briefly, and then began translating the delegate's English sentences into Spanish.

Let's unpick what she did that morning and itemise its components.

As the delegate spoke, Pinkney had to make sense of a message composed in one language while simultaneously constructing and articulating the same message in another tongue. The process required an extraordinary blend of sensory, motor and cognitive skills, all of which had to operate in unison. She did so continuously and in real time, without asking the speaker to slow down or clarify anything. She didn't stammer or pause. Nothing in our evolutionary history can have programmed Pinkney's brain for a task so peculiar and demanding. Executing it required versatility and nuance beyond the reach of the most powerful computers. It is a wonder that her brain, indeed any human brain, can do it at all.

Neuroscientists have explored language for decades and produced scores of studies on multilingual speakers. Yet understanding this process – simultaneous interpretation – is a much bigger scientific challenge. So much goes on in an interpreter's brain that it's hard even to know where to start. Recently, however, a handful of enthusiasts have taken up the challenge, and one region of the brain – the caudate nucleus – has already caught their attention.

The caudate isn't a specialist language area; neuroscientists know it for its role in processes like decision making and trust. It's like an orchestral conductor, coordinating activity across many brain regions to produce stunningly complex behaviours.

Which means the results of the interpretation studies appear to tie into one of the biggest ideas to emerge from neuroscience over the past decade or two. It's now clear that many of our most sophisticated abilities are made possible not by specialist brain areas dedicated to specific tasks, but by lightning-fast coordination between areas that control more general tasks, such as movement and hearing. Simultaneous interpretation, it seems, is yet another feat made possible by our networked brains.

§§§

Simultaneous interpretation often evokes a sense of drama. This may be because of its history: the creation of the League of Nations after World War I established the need for it, and use of the technique during the trials of senior Nazis at Nuremberg showcased its power. Doubts about accuracy lingered nonetheless; the UN Security Council didn't fully adopt simultaneous interpretation until the early 1970s. 'Until then they didn't trust the interpreters,' says Barbara Moser-Mercer, an interpreter and researcher at the University of Geneva. But now the two traditional capitals of the multilingual conference world – the UN offices in Geneva and New York – have been joined by Brussels, as the expanding European Union incorporates more and more languages. The current total is 24, and some meetings involve interpretation of every one.

Looking down over the delegates at the IMO, I was reminded of the view from a captain's bridge, or the gallery of a television studio. I had a feeling of control, a perverse reaction given that control is one thing interpreters lack. The words they utter and the speed at which they talk are determined by others. And even though Pinkney and Soliño had copies of some of the speeches that had been prepared for that morning, they had to be alive to humorous asides. Puns, sarcasm, irony and culture-specific jokes are an interpreter's nightmare. As one interpreter has noted in an academic article, 'Puns based on a single word with multiple meanings in the source language should generally not be attempted by interpreters, as the result will probably not be funny.' Quite.

Many of the delegates spoke in English, so the pressure on Anne Miles in the into-English booth down the hall was sporadic. Miles speaks French, German, Italian and Russian, and has been interpreting for 30 years. In between translating she told me about word order, another challenge that interpreters face daily. 'With German the 'nicht', the 'not', can come at the very end of the sentence. So you may be enthusing about something and then the speaker finally says 'nicht'. But if you're a German native you can hear the 'nicht' coming by the intonation.' Word order is a particular problem in fish meetings, which Miles said she dreads. In a long sentence about a particular variety of fish, and in a language where the noun – the name of the fish – comes towards the end, the interpreter is left guessing about the subject of the sentence until it's completed.

There's humour in these pitfalls, of course. Miles told me about an agricultural meeting at which delegates discussed frozen bull's semen; a French interpreter translated this as 'matelot congelés', or 'deep-frozen sailors'. And she shared an error of her own, produced when a delegate spoke of the need to settle something 'avant Milan' – 'before Milan', the city being the venue for a forthcoming meeting. Miles didn't know about the Milan summit, so said that the issue wasn't going to be settled for 'mille ans', or 'a thousand years'.

Some speakers talk too fast. 'There are various strategies. Some interpreters think it's best just to stop and just say the delegate is speaking too fast.' Miles herself doesn't find that useful because people have a natural pace, and someone asked to slow down is likely to pick up speed again. The alternative is to précis. 'You have to be quick on the uptake. It's not just language skills in this job, it's being quick-brained and learning fast.'

Challenges of this kind make simultaneous interpretation tiring, and explained why the two interpreters took it in turns to rest every half an hour. Watching by video is even worse. 'We don't like it at all,' Miles told me. Studies confirm that the process is more exhausting and stressful, probably because body language and facial expressions provide part of the message, and are harder to decipher when working remotely. 'You get fewer visual clues as to what's going on, even with a video link,' said Miles.

Then there's the tedium. Crisis talks in New York might be gripping, but the average politician, never mind the average technical expert on marine regulations, isn't likely to induce rapt attention for hours on end. The audience may slumber, but the interpreter must remain vigilant. As the meeting sailed on into a polyglot fog of procedural niceties and resolutions, each with sections and subsections, I realised how tiring this vigilance must be. Having nodded off in many a science conference – even once when chairing – I was in awe of the interpreters' fortitude.

§ § §

Moser-Mercer trained as an interpreter – she is fluent in German, English and French – before being sidetracked by neuroscience. 'I got very intrigued with what was going on in my brain while I was interpreting,' she says. 'I thought there has to be a way to find out.' When she arrived at the University of Geneva in 1987 there wasn't a way – the interpretation department was concerned with training, not research. So she set out to create one by collaborating with colleagues in the brain sciences.

'Language is one of the more complex human cognitive functions,' Narly Golestani, Group Leader of the university's Brain and Language Lab, tells me during a recent visit. 'There's been a lot of work on bilingualism. Interpretation goes one step beyond that because the two languages are active simultaneously. And not just in one modality, because you have perception and production at the same time. So the brain regions involved go to an extremely high level, beyond language.'

In Geneva, as in many other neuroscience labs, the tool of choice is functional magnetic resonance imaging (fMRI). Using fMRI, researchers can watch the brain as it performs a specific task; applied to interpretation, it has already revealed the network of brain areas that make the process possible. One of these is Broca's area, known for its role in language production and working memory, the function that allows us to maintain a grasp on what we're thinking and doing. The area is also linked with neighbouring regions that help control language production and comprehension. 'In interpretation, when a person hears something and has to translate and speak at the same time, there's very strong functional interplay between these regions,' says Golestani.

Many other regions also seem to be involved, and there are myriad connections between them. The complexity of this network deterred Moser-Mercer from tackling them all at once; unravelling the workings of each component would have been overwhelming. Instead the Geneva researchers treat each element as a black box, and focus on understanding how the boxes are linked and coordinated. 'Our research is about trying to understand the mechanisms that enable the interpreter to control these systems simultaneously,' says team member Alexis Hervais-Adelman.

Two regions in the striatum, the evolutionarily ancient core of the brain, have emerged as key to this executive management task: the caudate nucleus and the putamen. Neuroscientists already know that these structures play a role in other complex tasks, including learning and the planning and execution of movement. This means that there is no single brain centre devoted exclusively to the control of interpretation, say Hervais-Adelman and his colleagues. As with many other human behaviours studied using fMRI, it turns out that the feat is accomplished by multiple areas pitching in. And the brain areas that control the process are generalists, not specialists.

§§§

One of the triggers of this piece was a trivial conversation. Someone told me of a simultaneous interpreter so proficient that he could do a crossword while working. No name or date or place was mentioned, so I was sceptical. But just to check I contacted a few professional interpreters. One thought he might have heard a rumour; the others were dismissive. An urban myth, they said.

I ask Moser-Mercer if interpreters ever do anything else while interpreting. In a job dominated by women, she tells me, some knit – or used to when it was a more popular pastime. And you can see how a regular manual action might complement the cerebral activity of translation. But a crossword puzzle? Moser-Mercer hasn't tried it, but she tells me that under exceptional circumstances – a familiar topic, lucid speakers, etc – she thinks she could.

That such a feat might be possible suggests that interesting things are indeed happening in the brains of simultaneous interpreters. And there are other reasons for thinking that interpreters' brains have been shaped by their profession. They're good at ignoring themselves, for example. Under normal circumstances listening to your voice is essential to monitoring your speech. But interpreters have to concentrate on the word they're translating, so they learn to pay less attention to their own voice.

Some habits acquired in the workplace may carry over to the home. One way that experienced interpreters acquire speed is by learning to predict what speakers are about to say. 'I will always anticipate the end of a sentence, no matter who I'm talking to and whether or not I'm wearing a headset,' says Moser-Mercer. 'I will never wait for you to finish your sentence. Many of us interpreters know this from our spouses and kids. 'You never let me finish...' And it's true. We're always trying to jump in.'

Interpreters also have to be able to cope with stress and exercise self-control when working with difficult speakers. I read one review, based on questionnaires given to interpreters, which suggested that members of the profession are, as a consequence, highly strung, temperamental, touchy and prima donna-ish. Maybe. But I couldn't see it in Marisa, Carmen or Anne.

§ § §

A few years ago, the Geneva researchers asked 50 multilingual students to lie in a brain scanner and carry out a series of language exercises. In one subjects merely listened to a sentence and said nothing. Another involved the students repeating the sentence in the same language. The third was the most onerous: subjects were asked to repeat what they were hearing, this time translating it into another language.

In cognitive terms this seems like a big step up. Initially the students just had to listen, and then to repeat. Task three required them to think about meaning and how to translate it: to interpret simultaneously. But the scans didn't reveal any neural fireworks. 'There wasn't a huge amount of additional engagement,' says Hervais-Adelman. No extra activity in regions that handle comprehension or articulation, for example. 'It was just a handful of specific regions that were handling the extra load of the interpreting.' These included areas that control movement, such as the premotor cortex and the caudate. Interpretation, in other words, may be about managing specialised resources rather than adding substantially to them.

This idea remains unconfirmed, but the Geneva team added weight to it when they invited some of the same students back into the fMRI scanner a little over a year later. During that period 19 of the returnees had undergone a year of conference interpretation training, while the others had studied unrelated subjects. The brains of the trainee interpreters had changed, particularly parts of the right caudate, but not in the way you might expect – activity there lessened during the interpretation task. It is possible that the caudate had become a more efficient coordinator, or had learned how to farm out more of the task to other structures.

'It could be that as people become more experienced in simultaneous interpretation there's less need for the kind of controlled response provided by the caudate,' says David Green, a neuroscientist at University College London who was not involved in the Geneva work. 'The caudate plays a role in the control of all sorts of skilled actions. And there's other work showing that as people get more skilled at a task you get less activation of it.'

The story that is emerging from the Geneva work – that interpretation is about coordinating more specialised brain areas – seems to gel with interpreters' descriptions of how they work. To be really effective, for example, a simultaneous interpreter needs a repertoire of approaches. 'The process has to adapt to varying circumstances,' says Moser-Mercer, who still does 40 to 50 days of interpretation a year, mainly for UN agencies. 'There could be poor sound quality, or a speaker with an accent, or it might be a topic I don't know much about. For instance, I wouldn't interpret a fast speaker in the same way I would a slow one. It's a different set of strategies. If there isn't time to focus on each and every word that comes in you have to do a kind of intelligent sampling.' It may be that the flexible operation of the brain networks underpinning interpretation allows interpreters to optimise strategies for dealing with different types of speech. And different interpreters listening to the same material may use different strategies.

The results from the Geneva group also fit with a wider theme in neuroscience. When fMRI became widely available in 1990s, researchers rushed to identify the brain areas involved in almost every conceivable behaviour (including, yes, sex: several researchers have scanned the brains of subjects experiencing an orgasm). But on their own those data didn't prove terribly useful, partly because complex behaviours don't tend to be controlled by individual brain areas. Now the emphasis has shifted to understanding how different areas interact. Neuroscientists have learned that when we consider a potential purchase, for example, a network of areas that includes the prefrontal cortex and insula helps us decide whether the price is right. Interplay between another set of brain areas, including the entorhinal cortex and the hippocampus, helps store our memories of routes between places.

This more sophisticated understanding has been made possible in part by improvements in scanning technology. In the case of the caudate, activity there can now be distinguished from that in other parts of the basal ganglia, the larger brain area within which it is located. The finer-grained scans have revealed that the caudate is often involved in networks that regulate cognition and action, a role that puts it at the heart of an extraordinarily diverse range of behaviours. As a team of British researchers noted in a 2008 review, studies have shown that the caudate helps control everything from 'a rat's decision to press a lever to a human's decision about how much to trust a partner in a financial exchange'.

One of the review's authors was John Parkinson of Bangor University in Wales. I ask him if he would have predicted that the caudate would be involved in simultaneous interpretation. He says that at first he wouldn't have. 'The caudate is involved in the intentionality of an action, in its goal-directedness. Not so much in carrying it out but in why you're doing it.' Then he thought about what interpreters do. Computers translate by rote, often with risible results. Humans have to think about meaning and intent. 'The interpreter must actually try to identify what the message is and translate that,' says Parkinson. He agrees that the involvement of the caudate makes sense.

Given that the Geneva research is based partly in a department tasked with training interpreters, it's natural to wonder if their scientific findings might eventually find a direct practical application. Moser-Mercer and her colleagues are careful to avoid extravagant claims, and rule out suggestions that brain scanners might be used to assess progress or select candidates with an aptitude for interpreting. But even if studying simultaneous interpretation doesn't lead to immediate applications, it has already extended our knowledge of the neural pathways that link thinking with doing, and in the future it may help neuroscientists gain an even deeper understanding of the networked brain. The Geneva team wants next to explore the idea that some high-level aspects of cognition have evolved from evolutionarily older and simpler behaviours. The brain, they suggest, builds its complex cognitive repertoire upon on a lower level of what they call 'essential' processes, such as movement or feeding. 'This would be a very efficient way to do things,' Moser-Mercer and her colleagues tell me in an email. 'It makes sense for the brain to evolve by reusing or by adapting its processors for multiple tasks, and it makes sense to wire the cognitive components of control directly into the system that will be responsible for effecting the behaviour.' Simultaneous interpreting, with its close back-and-forth relationship between cognition and action, may be an ideal test bed for such thinking.

This story was first published on 18 November 2014 by Wellcome on mosaicscience.com

How should we deal with dark winters?

■ Linda Geddes

The inhabitants of Rjukan in southern Norway have a complex relationship with the sun. 'More than other places I've lived, they like to talk about the sun: when it's coming back, if it's a long time since they've seen the sun,' says artist Martin Andersen. 'They're a little obsessed with it.' Possibly, he speculates, it's because for approximately half the year, you can see the sunlight shining high up on the north wall of the valley: 'It is very close, but you can't touch it,' he says. As autumn wears on, the light moves higher up the wall each day, like a calendar marking off the dates to the winter solstice. And then as January, February and March progress, the sunlight slowly starts to inch its way back down again.

Rjukan was built between 1905 and 1916, after an entrepreneur called Sam Eyde bought the local waterfall (known as the smoking waterfall) and constructed a hydroelectric power plant there. Factories producing artificial fertiliser followed. But the managers of these worried that their staff weren't getting enough sun – and eventually they constructed a cable car in order to give them access to it.

When Martin moved to Rjukan in August 2002, he was simply looking for a temporary place to settle with his young family that was close to his parents' house and where he could earn some money. He was drawn to the three-dimensionality of the place: a town of 3,000, in the cleft between two towering mountains – the first seriously high ground you reach as you travel west of Oslo.

But the departing sun left Martin feeling gloomy and lethargic. It still rose and set each day, and provided some daylight – unlike in the far north of Norway, where it is dark for months at a time – but the sun never climbed high enough for the people of Rjukan to actually see it or feel its warming rays directly on their skin.

As summer turned to autumn, Martin found himself pushing his two-year-old daughter's buggy further and further down the valley each day, chasing the vanishing sunlight. 'I felt it very physically; I didn't want to be in the shade,' says Martin, who runs a vintage shop in Rjukan town centre. If only someone could find a way of reflecting some sunlight down into the town, he thought. Most people living at temperate latitudes will be familiar with Martin's sense of dismay at autumn's dwindling light. Few would have been driven to build giant mirrors above their town to fix it.

What is it about the flat, gloomy greyness of winter that seems to penetrate our skin and dampen our spirits, at least at higher latitudes? The idea that our physical and mental health varies with the seasons and sunlight goes back a long way. The Yellow Emperor's Classic of Medicine, a treatise on health and disease that's estimated to have been written around 300 BCE, describes how the seasons affect all living things and suggests that during winter – a time of conservation and storage – one should 'retire early and get up with the sunrise... Desires and mental activity should be kept quiet and subdued, as if keeping a happy secret.' And in his Treatise on Insanity, published in 1806, the French physician Philippe Pinel noted a mental deterioration in some of his psychiatric patients 'when the cold weather of December and January set in'.

Today, this mild form of malaise is often called the winter blues. And for a minority of people who suffer from seasonal affective disorder (SAD), winter is quite literally depressing. First described in the 1980s, the syndrome is characterised by recurrent depressions that occur annually at the same time each year. Most psychiatrists regard SAD as being a subclass of generalised depression or, in a smaller proportion of cases, bipolar disorder.

Seasonality is reported by approximately 10 to 20 per cent of people with depression and 15 to 22 per cent of those with bipolar disorder. 'People often don't realise that there is a continuum between the winter blues – which is a milder form of feeling down, [sleepier and less energetic] – and when this is combined with a major depression,' says Anna Wirz-Justice, an emeritus professor of psychiatric neurobiology at the Centre for Chronobiology in Basel, Switzerland. Even healthy people who have no seasonal problems seem to experience this low-amplitude change over the year, with worse mood and energy during autumn and winter and an improvement in spring and summer, she says.

Why should darker months trigger this tiredness and low mood in so many people? There are several theories, none of them definitive, but most relate to the circadian clock – the roughly 24-hour oscillation in our behaviour and biology that influences when we feel hungry, sleepy or active. This is no surprise given that the symptoms of the winter blues seem to be associated with shortening days and longer nights, and that bright light seems to have an antidepressive effect. One idea is that some people's eyes are less sensitive to light, so once light levels fall below a certain threshold, they struggle to synchronise their circadian clock with the outside world. Another is that some people produce more of a hormone called melatonin during winter than in summer – just like certain other mammals that show strong seasonal patterns in their behaviour.

However, the leading theory is the 'phase-shift hypothesis': the idea that shortened days cause the timing of our circadian rhythms to fall out of sync with the actual time of day, because of a delay in the release of melatonin. Levels of this hormone usually rise at night in response to darkness, helping us to feel sleepy, and are suppressed by the bright light of morning. 'If someone's biological clock is running slow and that melatonin rhythm hasn't fallen, then their clock is telling them to keep on sleeping even though their alarm may be going off and life is demanding that they wake up,' says Kelly Rohan, a professor of psychology at the University of Vermont. Precisely why this should trigger feelings of depression is still unclear. One idea is that this tiredness could then have unhealthy knock-on effects. If you're having negative thoughts about how tired you are, this could trigger a sad mood, loss of interest in food, and other symptoms that could cascade on top of that.

However, recent insights into how birds and small mammals respond to changes in day length have prompted an alternative explanation. According to Daniel Kripke, an emeritus professor of psychiatry at the University of California, San Diego, when melatonin strikes a region of the brain called the hypothalamus, this alters the synthesis of another hormone – active thyroid hormone – that regulates all sorts of behaviours and bodily processes.

When dawn comes later in the winter, the end of melatonin secretion drifts later, says Kripke. From animal studies, it appears that high melatonin levels just after the time an animal wakes up strongly suppress the making of active thyroid hormone – and lowering thyroid levels in the brain can cause changes in mood, appetite and energy. For instance, thyroid hormone is known to influence serotonin, a neurotransmitter that regulates mood. Several studies have shown that levels of brain serotonin in humans are at their lowest in the winter and highest in the summer. In 2016, scientists in Canada discovered that people with severe SAD show greater seasonal changes in a protein that terminates the action of serotonin than others with no or less severe symptoms, suggesting that the condition and the neurotransmitter are linked.

It's possible that many of these mechanisms are at work, even if the precise relationships haven't been fully teased apart yet. But regardless of what causes winter depression, bright light – particularly when delivered in the early morning – seems to reverse the symptoms.

It was a bookkeeper called Oscar Kittilsen who first came up with the idea of erecting large rotatable mirrors on the northern side of the valley, where they would be able to 'first collect the sunlight and then spread it like a headlamp beam over the town of Rjukan and its merry inhabitants'.

A month later, on 28 November 1913, a newspaper story described Sam Eyde pushing the same idea, although it was another hundred years before it was realised. Instead, in 1928 Norsk Hydro erected a cable car as a gift to the townspeople, so that they could get high enough to soak up some sunlight in winter. Instead of bringing the sun to the people, the people would be brought to the sunshine.

Martin Andersen didn't know all of this. But after receiving a small grant from the local council to develop the idea, he learned about this history and started to develop some concrete plans. These involved a heliostat: a mirror mounted in such a way that it turns to keep track of the sun while continually reflecting its light down towards a set target – in this case, Rjukan town square.

The three mirrors, each measuring 17 m², stand proud upon the mountainside above the town. In January, the sun is only high enough to bring light to the square for two hours per day, from midday until 2pm, but the beam produced by the mirrors is golden and welcoming. Stepping into the sunlight after hours in permanent shade, I become aware of just how much it shapes our perception of the world. Suddenly, things seem more three-dimensional; I feel transformed into one of those 'merry inhabitants' that Kittilsen imagined. When I leave the sunlight, Rjukan feels a flatter, greyer place.

As far back as the sixth century, historians were describing seasonal peaks of joy and sorrow among Scandinavians, brought about by the continuous daylight of summer and its almost complete absence in winter.

Three hundred and fifty miles south of Rjukan, and at roughly the same latitude as Edinburgh, Moscow and Vancouver, lies Malmö in southern Sweden. In Sweden, an estimated 8 per cent of the population suffer from SAD, with a further 11 per cent said to suffer the winter blues.

In early January, the sun rises at around 8.30am and sets just before 4pm. For Anna Odder Milstam, an English and Swedish teacher, this means getting up and arriving at work before dawn for several months of the year. 'During the winter, we just feel so tired,' she says. 'The children struggle with it as well. They are less alert and less active at this time of year.'

Anna picks me up from my city-centre hotel at 7.45am. It's early January and still dark, but as dawn begins to break it reveals a leaden sky and the threat of snow. I ask if she's a winter person and she visibly shudders. 'No, I am not,' she replies stiffly. 'I like the sun.'

Lindeborg School, where Anna teaches, caters for approximately 700 pupils, ranging from preschool age through to 16. Since there's little the school can do about its high latitude and brooding climate, the local authority is instead trying recreate the psychological effects of sunshine on its pupils artificially.

When I walk into Anna's classroom at 8.50am, my eyes instinctively crinkle, and I feel myself recoiling. It's as if someone has thrown open the curtains on a darkened bedroom. Yet as my eyes adjust to the bright light, I see the curtains in this classroom are firmly closed. In front of me sit a class of 14-year-olds at evenly spaced desks, watching my reaction with mild amusement. They're part of an experiment investigating whether artificial lighting can improve their alertness and sleep, and ultimately result in improved grades.

'We can all feel that if we're not very alert at school or work, we don't perform at our top level,' says Olle Strandberg, a developer at Malmö's Department of Internal Services, which is leading the project. 'So if there is any possibility of waking the students up during the wintertime, we're keen to take it.'

Since October 2015, Anna's classroom has been fitted with ceiling lights that change in colour and intensity to simulate being outside on a bright day in springtime. Developed by a company called BrainLit, the ultimate goal is to create a system that is tailored to the individual, monitoring the type of light they've been exposed to through the course of a day and then adjusting the lights to optimise their health and productivity.

When Anna's pupils enter the classroom at 8.10am, the lights are a bright bluish-white to wake them up. They then grow gradually more intense as the morning progresses, dimming slightly in the run up to lunch to ease the transition to the gloomier light outside. Immediately after lunch the classroom is intense whitish-blue again – 'to combat the post-lunch coma,' jokes Strandberg – but then the lights gradually dim and become more yellow as the afternoon progresses.

Bright light in the morning suppresses any residual melatonin that could be making us sleepy, and provides a signal to the brain's master clock that keeps it synchronised with the 24-hour cycle of light and dark. The idea is it therefore strengthens our internal rhythms, so that when night comes around again, we start to feel sleepy at the correct time.

Already, there's some preliminary evidence that it's having an effect on the pupils' sleep. In a small pilot study, 14 pupils from Anna's class and 14 from a neighbouring class that doesn't have the lighting system were given Jawbone activity trackers and asked to keep sleep diaries for two weeks. During the second week, significant differences started to emerge between the two groups in terms of their sleep, with Anna's pupils waking up fewer times during the night and spending a greater proportion of their time in bed asleep.

No one knows whether the lighting system is affecting the students' exam scores, or even how to measure that. But it might. Besides suppressing melatonin and warding off any residual sleepiness, recent studies suggest that bright light acts as a stimulant to the brain. Gilles Vandewalle and colleagues at the University of Liège in Belgium asked volunteers to perform various tasks in a brain scanner while exposing them to pulses of bright white light or no light. After exposure to white light, the brain was in a more active state in those areas that were involved in the task. Although they didn't measure the volunteers' test performances directly, if you are able to recruit a greater brain response, then your performance is likely to be better: you will be faster or more accurate, Vandewalle says.

Anna agrees. Anecdotally, she reports that her students are more alert. 'They've expressed that they feel more able to concentrate and they are more focused,' she says. 'I also look forward to going into my classroom in the morning, because I've noticed that I feel better when I go in there – more awake.'

Of course, the idea of using light to counter the winter blues is nothing new. SAD lamps are a mainstay of treatment for winter depression, and in Sweden, which was a vigorous early adopter of light therapy, clinics often went one step further: dressing patients in all-white clothes and sending them into white rooms filled with bright light.

Baba Pendse, a Malmö-based psychiatrist, recalls visiting one of these early light rooms in Stockholm in the late 1980s: 'I remember that after being in there for some time, we all started to get very lively,' he says. In 1992, he opened a light therapy clinic in Lund, and another in neighbouring Malmö a few years later, which still exist today.

Sitting in the Malmö light room with Pendse brings back memories of sunny cafés at the top of ski slopes: the brightness elicits the same sense of elation. The room contains 12 white chairs and footstalls, each draped in a white towel and clustered around a white coffee table stacked with white cups, napkins and sugar cubes. The only non-white object in the room is a jar of instant coffee granules. It's warm, and the lights emit a very faint hum. Around 100 people diagnosed with SAD use the light room each winter, initially booking in for 10 two-hour sessions in the early morning over the course of two weeks. Pendse offers his patients the choice of group light therapy or taking antidepressants to combat their depression. 'But unlike antidepressants, with light therapy you get an almost immediate effect,' he says.

In recent years, light therapy has experienced something of a backlash in Sweden, and Malmö's clinic is one of only a handful that remain. In part, this was a response to a 2007 study by the Swedish Council on Technology Assessment in Health Care, which reviewed the available evidence and concluded that 'although treatment in light therapy rooms is well established in Sweden, no satisfactory, controlled studies have been published on the subject'. They said the value of therapy with a light box for SAD 'can be neither confirmed nor dismissed', which, while inconclusive, was interpreted by some as 'light therapy has no effect'.

Pendse shakes his head telling me this. Conducting gold-standard, randomised placebo-controlled studies of light therapy is difficult, he says, because 'what do you use as a placebo?'

Even so, there's some evidence that light therapy may have a similar effect on the brain to many antidepressants. In a study published in 2016, 11 patients with SAD treated with two weeks of light therapy saw plunging levels of serotonin transporter binding – a measure of how quickly serotonin's activity is curtailed. Their levels became similar to those seen during summertime.

There is other evidence, besides. At the back of our eyes, an unusual type of photoreceptor has been found that seems to help synchronise our circadian rhythms to the 24-hour cycle of light and dark. These cells, called ipRGCs, are particularly sensitive to blue light, connect to a number of different brain areas, and seem to feed into our circadian clock, our sleep centres and even some mood-regulating areas.

Through these cells, bright light seems to affect our mood and alertness in several ways – suppression of melatonin and synchronisation of the circadian clock, for instance – but researchers believe they have another, more direct impact on mood. Mouse studies have found that bright light at inappropriate times of day leads to depression-like behaviours (the mice became less interested in sugar and quickly gave up when challenged with a forced swimming test – a common measure of despair in mice). But this doesn't happen in mice genetically engineered to lack ipRGCs.

Vandewalle's lab meanwhile has discovered that the brain areas responsible for processing emotion light up more strongly in response to blue light, and has found an abnormal response to blue light in the hypothalamus of SAD sufferers during the winter months. 'People with winter depression tend to sleep more, eat more and to be demotivated. The hypothalamus is implicated in all these areas, so it may be an important region for the impact of light on the brain,' he says.

Not everyone in Rjukan has welcomed the sun mirrors with open arms. Many of the locals I spoke to dismissed them as a tourist gimmick – though all admitted they were good for business. On the day I visited, the town was blessed with clear blue skies and a golden shaft of light descending from the mirrors, yet few people lingered in the town square. In fact, of the people I spoke to, it was recent immigrants to Rjukan who seemed most appreciative of the mirrors.

Martin Andersen admits to having got used to the lack of light over time. 'I don't find it so bad anymore,' he says. It's as though the people who've been brought up in this uniquely shady place, or who have chosen to stay, have grown immune to the normal thirst for sunlight.

This is certainly the case in another Norwegian settlement: Tromsø. One of the world's most northerly cities, it is some 400 km north of the Arctic Circle. Winter in Tromsø is dark – the sun doesn't even rise above the horizon between 21 November and 21 January. Yet strangely, despite its high latitude, studies have found no difference between rates of mental distress in winter and summer.

One suggestion is that this apparent resistance to winter depression is genetic. Iceland similarly seems to buck the trend for SAD: it has a reported prevalence of 3.8 per cent, which is lower than that of many countries farther south. And among Canadians of Icelandic descent living in a region of Canada called Manitoba, the prevalence of SAD is approximately half that of non-Icelandic Canadians living in the same place.

However, an alternative explanation for this apparent resilience in the face of darkness is culture. 'To put it brutally and brief: it seems like there are two sorts of people who come up here,' says Joar Vit[tersø, a happiness researcher at the University of Tromsø. 'One group tries to get another kind of work back down south as soon as possible; the other group remains.'

Ane-Marie Hektoen grew up in Lillehammer in southern Norway, but moved to Tromsø 33 years ago with her husband, who grew up in the north. 'At first I found the darkness very depressing; I was unprepared for it, and after a few years I needed to get a light box in order to overcome some of the difficulties,' she says. 'But over time, I have changed my attitude to the dark period. People living here see it as a cosy time. In the south the winter is something that you have to plough through, but up here people appreciate the very different kind of light you get at this time of year.'

Stepping into Ane-Marie's house is like being transported into a fairy-tale version of winter. There are few overhead lights, and those that do exist drip with crystals, which bounce the light around. The breakfast table is set with candles, and the interior is furnished in pastel pinks, blues and white, echoing the soft colours of the snow and the winter sky outside. It is the epitome of kos or koselig – the Norwegian version of hygge, the Danish feeling of warmth and cosiness.

The period between 21 November and 21 January in Tromsø is known as the polar night, or dark period, but for at least several hours a day it isn't strictly speaking dark, but more of a soft twilight. Even when true darkness does descend, people stay active. One afternoon I hire a pair of cross-country skis and set off down one of the street-lit tracks that criss-cross the edge of the city. Despite the darkness, I encounter people taking dogs for walks on skis, a man running with a head torch, and countless children having fun on sledges. I stop at a park and marvel at a children's playground lit up by floodlights. 'Do children climb here in winter?' I ask a young woman, who is struggling to pull on a pair of ice skates. 'Of course,' she answers, in perfect English. 'It's why we have floodlights. If we didn't, we'd never get anything done.'

During 2014–15, a psychologist from Stanford University called Kari Leibowitz spent ten months in Tromsø trying to figure out how people cope during the cold, dark winters. Together with Vittersø, she devised a 'winter mindset questionnaire' to assess people's attitudes to winter in Tromsø, Svalbard and the Oslo area. The farther north they went, the more positive people's mindsets towards winter were, she tells me. 'In the south, people didn't like winter nearly as much. But across the board, liking winter was associated with greater life satisfaction and being willing to undertake challenges that lead to greater personal growth.'

It sounds dismissively simple, but adopting a more positive attitude really might help to ward off the winter blues. Kelly Rohan recently published a clinical trial comparing cognitive behavioural therapy (CBT) to light therapy in the treatment of SAD, and found them comparable during the first year of treatment. CBT involves learning to identify patterns and errors in one's way of thinking and challenging them. In the case of SAD, that could be rephrasing thoughts such as 'I hate winter' to 'I prefer summer to winter', or 'I can't do anything in winter' to 'It's harder for me to do things in winter, but if I plan and put in effort I can'.

It also involves finding activities that a person is willing to do in winter, to pull them out of hibernation mode. 'I don't argue that there isn't a strong physiological component to seasonal depression, which is tied to the light–dark cycle,' says Rohan. 'But I do argue that the person has some control over how they respond to and cope with that. You can change your thinking and behaviour to feel a bit better at this time of year.'

This story was first published on 12 January 2016
by Wellcome on mosaicscience.com

Smartphones won't* make your kids dumb (*Probably)

■ Olivia Solon

Jessica's tiny fingers dart around the iPad, swiping through photos to get to a particularly entertaining video: a 12-second clip of her dancing clumsily to Beyoncé's Single Ladies. The 18-month-old taps 'play' and emits a squeal of delight.

After watching the video twice, she navigates back to the home screen and opens up the YouTube app to watch an episode of the colourful animation Billy Bam Bam. Halfway through, she moves onto a Yo Gabba Gabba! game, which involves anthropomorphised fruits making their way into a character's belly.

When Jessica's mum, Sandy, tries to take away the iPad, there's a tantrum that threatens to go nuclear: wobbly lip, tears, hands balled into fists and a high-pitched wail. 'She does this a lot,' says Sandy. 'She seems to prefer the iPad to everything else. Sometimes it's the only thing that will keep her quiet,' she adds, frantically waving a pink fluffy unicorn in an attempt to appease her daughter.

Like many parents, she's worried about her child's obsession with screens. She wants to know which activities are best, and how much time spent on screens is too much.

It's six years since the launch of the iPad and, with it, the rebirth of tablet computers. The academic research simply hasn't been able to catch up, which means it's hard to know the long-term impact on young brains of being exposed to tablets and smartphones.

The concern among some experts is that these devices, if used in particular ways, could be changing children's brains for the worse – potentially affecting their attention, motor control, language skills and eyesight, especially in under-fives, for whom so much brain development is taking place.

Technology companies and app developers are throwing their marketing prowess at the problem, slapping words like 'educational' and 'e-learning' on their products, often without any scientific basis. So what are parents to do?

§ § §

People have always feared new media. Almost 2,500 years ago Socrates was decrying the spread of written language, arguing that it would erode memory and knowledge. In the 15th century it was the printing press that brought about moral panic. Benedictine monks, who profited from hand-copying reading materials, petitioned against the mechanised printers, saying: 'They shamelessly print, at negligible cost, material which may, alas, inflame impressionable youths.'

When radio arrived, it too was deemed a menace, blamed for distracting children from their homework. A 1936 article in Gramophone magazine reported that youngsters had 'developed the habit of dividing attention between the humdrum preparation of their school assignments and the compelling excitement of the loudspeaker'.

Few technologies, however, have invaded our lives – and those of our children – as stealthily as the mobile computer, most commonly the smartphone or tablet. These devices are the right size for little hands to handle them, and the touchscreens easy for tiny fingers to manipulate. Plus there's so much you can do on these devices: watch videos, play games, draw pictures and talk to relatives thousands of miles away.

Sometimes the iPad is the only thing that will keep her quiet

In 2011, a year after the iPad launched, just 10 per cent of US children under the age of two were found to have used tablets or smartphones, but by 2013 that figure had nearly quadrupled. A 2015 study in France found that 58 per cent of under-twos had used a tablet or mobile phone.

There's little clarity around the consequences of long-term use of such devices. The American Academy of Pediatrics (AAP) has erred on the side of caution, recommending absolutely no screen time for children under the age of two, and a two-hour daily limit for those older. These restrictions simply don't tally with how many people are integrating these devices into their children's lives, nor do they reflect the fact that some interactions with screens might actually be beneficial.

'If your child is under two and is exposed to a screen it's not going to be toxic to their brain: they won't be turned into idiots,' says Michael Rich, Associate Professor of Pediatrics at Harvard Medical School and an AAP member. 'But there are potential downsides... and parents need to make a series of risk–benefit analyses.' The AAP is now in the process of revising its guidelines, and they are due to be published in late 2016.

So why don't we know more about the risks of children using screens? There's a fundamental problem at the basis of all the research in this area – what do we even mean by 'screen time'?

Firstly, it's important to distinguish between types of screen: do we mean a television screen, a tablet, a smartphone or an e-reader? Secondly, the nature of the content matters: is it an interactive drawing game, an e-book, a Skype call with Grandma or a stream of Netflix Kids videos? Thirdly, there's the context: is there a caregiver in the room talking to the child as they interact with the screen or are they left on their own?

To date, we have comprehensive research about children and television exposure, but we don't yet know how much of it applies to interactive screens like tablets and smartphones.

There are a few things we do know. Most child development experts agree that while passive screen time – such as putting your child in front of a device for a Peppa Pig marathon – might be entertaining, it isn't going to provide a rich learning experience. In this case, it doesn't make a difference whether they're watching on TV or a tablet: the experience is broadly the same.

Having a video or TV on when a child is doing something else can distract them from play and learning, negatively affecting their development. Hours of background TV has also been found to reduce child–parent interaction, which has an adverse impact on language development. This displacement is a big concern: if kids are left with screen-based babysitters then they are not interacting with caregivers and the physical world. There are only so many hours in a day, and the time spent with screens comes at the expense of other, potentially better, activities.

Under-threes, in particular, need a balance of activities, including instructed play, exploring the natural environment, manipulating physical toys and socialising with other children and grown-ups. The rise in screen use means less of all of these things. 'Parents need to think strategically,' says paediatrician Dimitri Christakis, Director of the Center for Child Health, Behaviour and Development at the Seattle Children's Research Institute. 'If your child has 12 hours awake and two of those are spent eating, how will you allocate the rest of the time?'

The problem is that tablets are extremely appealing to children and adults alike. Thanks to their design, versatility and intuitive interfaces, tablets are a perfect way for children to draw, solve puzzles and be entertained on the move. Combine that with marketing efforts of digital media companies and app developers – whose measure of success tends to be the amount of time people are glued to their creation – and you have a toy that's difficult to prise out of tiny hands.

Many apps are designed to be stimulus-driven, with exciting audiovisual rewards for completing tasks. Christakis refers to this as the 'I did it!' response, which triggers the reward pathway in the brain. 'The delight a child gets from touching a screen and making something happen is both edifying and potentially addictive,' he says.

Because of this, tablets and smartphones make for excellent pacifiers, particularly on long plane journeys and in restaurants. 'The device itself is seen as a pleasurable source of comfort and parents play into that,' says Christakis.

'It's pretty common,' says Jenny Radesky, Assistant Professor of Pediatrics at the University of Michigan. 'It becomes the go-to, easiest tool the parent is using.' Although helpful in the short term, it's important for young children to be able to develop internal mechanisms of self-regulation, whether that's learning without constant rewards or being able to sit patiently without constant digital stimulation.

Christakis says that, anecdotally, he and others are starting to see younger and younger patients using these devices compulsively. 'We know there's such a thing as problematic internet use in older children and adolescents, and it stands to reason that the same would happen with infants,' he says. And he's doing research to find out more about this.

In Seattle's Center for Integrative Brain Research, a cluster of tiny pink mouse pups wriggle in a mass behind their mother. The rodent family home is a sawdust-filled clear plastic container, one of hundreds stacked up in a rotating system of shelves. These are the 'control' mice used by Christakis and neuroscientist Nino Ramirez, and their team, trying to understand the impact on young brains of being exposed to fast-paced media.

Across the corridor an experiment is underway. One of the mouse containers is surrounded by bright lights and speakers. For 42 days, six hours every day, baby mice are exposed to the high-octane soundtrack of Cartoon Network shows accompanied by synchronised flashing lights in red, blue and green. The apparatus has been designed to find out what happens to the rodents' brains when they are overstimulated by media during a critical window for their development.

The results are startling. 'Overstimulating them as babies primes [them] to become hyperactive for the rest of their life,' says Ramirez. The overstimulated mice take more risks and find it harder to learn and stay attentive. They get confused by objects they've seen before, for example, and find it more difficult to navigate through a maze. When given the option to dose themselves with cocaine, the overstimulated mice were much more prone to addiction than the control group. These behavioural changes are matched by changes in the mice's brains.

The theory is that the same applies to children: overstimulating them with media – particularly in an age of tablets with endlessly streaming, hard-to-ration videos and flashy interactive games – may cause an imbalance in part of the cerebral cortex called the basal ganglia. It's this part of the brain that allows us to pay attention to critical tasks and ignore distractions.

Such overstimulation could lead to problems in later life, particularly with focus, memory and impulsivity. 'It seems that you can overstimulate young brains to the point that day-to-day life won't excite to the same extent,' says Ramirez.

Before we trigger mass panic about a generation of hyperactive, inattentive, cocaine-using post-millennials, it's important to note that these experiments have attracted criticism for a number of reasons. Six hours of any activity per day is a huge amount of time, particularly when it involves nocturnal mammals like mice (although the researchers say the mice show no signs of stress). Furthermore, Christakis, Ramirez and colleagues don't actually show the mice a real screen with any meaningful content – it is just a flashing proxy for a screen.

The rodent research being carried out in Seattle is unique in its scope and approach, which explains why it's frequently used as evidence of the evils of screen time. While mouse models are by no means perfect, they are useful for studying the underlying mechanisms relating to basic cognitive processes, which are fairly constant throughout mammals.

As mice have relatively short lifespans, it's possible to examine developmental trajectories over much shorter timeframes and get real insight into what's going on inside their brains. All of this can be done in a controlled environment that simply would not be possible with human subjects.

If, as suggested, cognitive development is affected by exposure to media, then this kind of research could inform the types of screen-based interactions we allow young children to have. Should parents be concerned? 'They should be vigilant and careful about the amount of time and the content their children have access to,' says Christakis.

Although it's a challenge to conduct controlled experiments with babies, it is possible to observe what happens with children 'in the wild'. From this, we can draw possible links to their habits with mobile devices.

In California, Maria Liu heads up the Myopia Control Clinic at UC Berkeley's School of Optometry. She's seen a sharp increase in young children with myopia (shortsightedness). 'It's increasing at an alarming rate worldwide and a well-accepted contributing factor is the early introduction of handheld devices to kids.'

In our early years, our eyeballs are very adaptive and plastic, so spending lots of time focusing on objects close-up will make the eyes more likely to be near-sighted. 'The eyeball will grow longer to compensate for the prolonged near stress,' Liu says. She doesn't have any evidence-based recommendations for a time limit on use of devices, but says 'frequent breaks from near work' are very important.

Tablets and smartphones are typically viewed much closer to the face than things like televisions or desktop computers. Although books are also read up close, studies have shown that children tend to hold them further away than they do screens.

The other problematic aspect of screens is that they have been shown to disrupt sleep. The blue light emitted by the super-sharp displays can interfere with our natural bodily rhythms, preventing melatonin, an important sleep hormone, from being released. This in turn can lead to sleep impairments in adults and children alike. Sandy says that if Jessica uses the tablet before bed she gets 'noticeably riled up'. So, she says, they try to use books instead. This issue is why the latest version of Apple's software for iPads and iPhones comes with 'Night Shift', which automatically swaps the blueish light for a warmer hue before bedtime.

§§§

In London, Max, who is 12 months old, is sitting on his mother Helen's lap in a small, darkened room. On his head is a rubbery cap covered in electrodes. They are measuring the electrical activity in his brain as he looks at physical objects and at digital representations of those objects on an iPad screen. On each of Max's ankles is a smartwatch of sorts, one measuring his movements and the other his heart rate. The cap uses electroencephalography (EEG) to record his brain's electrical activity, to understand whether real and virtual objects trigger different brain responses and how that relates to subsequent learning.

The experiment is part of the TABLET project in the Babylab at Birkbeck, University of London. It's the first ever scientific study investigating how children aged six months to three years are using touchscreen devices and how this influences their cognitive, brain and social development.

In a second experiment, Max sits in a curtained-off booth facing a screen that displays a 15-minute loop of video that includes trippy abstract animations and sounds, as well as still pictures and videos starring PhD students as stand-in children's TV presenters. He's completely mesmerised, and his eyes dart from object to object on the screen. Eye-tracking cameras capture the dance of his gaze, and outside the booth research fellow Celeste Chung keeps track of how his eye movements match up with the items on screen.

'All the child is doing is looking at the screen, but their gaze behaviour tells us about their learning and anticipations,' says Tim Smith, a cognitive scientist who heads up the Babylab.

The team is trying to understand how easily Max, and dozens of other babies like him, can focus attention and block out distractions when working on a particular task. In one of the tests, an object appears at the centre of the screen and then a second object appears, near the edge of the screen, shortly after. In order to look at the second object, the child needs to disengage from the central one, which requires self-control. This is a very important measure of executive function, the brain's 'air traffic control system', which helps a child analyse tasks, break them into steps and focus on them until they are done – a key predictor of success in later life.

Like Christakis, Smith is interested in finding out whether there really is a link between the reward learning found in many apps and a child's ability to focus. 'We might find that if tablets are being used for a lot of reward learning and the child becomes driven by an external stimulus then they might develop an impairment in executive function because they aren't used to controlling their own attention,' he says.

Smith isn't entirely convinced by the mouse model used by Christakis and Ramirez in Seattle, although he agrees that their six hours of media stimulation a day could be reflective of a small number of children's home environments where there are multiple devices and televisions that can contribute to sensory overload. 'Some of the parents in our study are reporting three hours of tablet use a day [for their children],' says Smith. 'That is a large proportion of their waking hours using a screen that doesn't conform to the laws of physical reality.'

As for the effects on language and motor development, he hypothesises that there could be displacement going on. 'The technology may be used as a nanny in place of face-to-face learning. Babies always learn better from people, but we don't always have time.' Devices like iPads may give lots of stimulation but lack the nuanced real-time social feedback that helps develop language, says Smith. Similarly, tablets and phones may make children dexterous at fine motor control with all the tapping and swiping, but they may have less motivation to get up and explore the world around them.

After around an hour of assessment, Max's patience for screen-touching, eye-tracking, brain-monitoring and other distractions from his busy schedule of rampaging around and eating bread sticks wanes. He starts to grizzle and wriggle and claw at the EEG cap. These movements corrupt the brain activity data. 'That's the interesting challenge with infants,' says Smith. 'They're completely non-compliant to instructions.'

What about the educational potential of devices? There are thousands of apps, e-books and videos purporting to have educational value for children, yet very few have been able to support this claim with solid research.

'The app marketplace is a digital Wild West,' says Michael Levine, Executive Director of the Joan Ganz Cooney Center in New York, which has analysed hundreds of children's literacy apps in a series of reports. 'Most of the apps labelled as educational provide no research-based advice or guidance... Less than 10 per cent of the apps we looked at had any stated evidence of efficacy [in the descriptions in the app store].'

Unintentionally, some interactive 'enhancements' to stories (such as animations, sounds and features that let kids tap and swipe) might actually be decreasing the overall educational value. While enhancements might appear to be engaging children, they could, in fact, be distracting them from the educational content.

This idea was put to the test by Adriana Bus and colleagues, at Leiden University in the Netherlands, who tracked children's eyes while they read interactive e-books. They found that when there were animated parts of the picture not directly relevant to the narrative – for example, trees moving in the wind in the background – the children's eyes were diverted to those points of motion rather than taking in the story. Relevant animations, on the other hand, can be beneficial, particularly for children who struggle with language and reading comprehension.

Even if apps are found to have educational value, toddlers still learn better from experiences in the real world than they do from equivalent two-dimensional representations on screen. Studies in the US have shown that when dealing with visual–spatial problems, such as finding hidden objects or solving puzzles, toddlers (under around 30 months) perform much better when the problem is presented in real life rather than on screen.

'It is thought that the cognitive load of transferring information from two dimensions... to three dimensions... is too great for children prior to age 30 months,' write Jenny Radesky and her colleague Barry Zuckerman in their study of digital play. Children this young are still developing the ability to choose what to pay attention to and what to ignore, and they have trouble generalising from symbolic representations to the real world.

Preschool-aged children need to interact with actual physical objects in order to develop their parietal cortex, which controls visual–spatial processing and helps develop maths and science skills in later life. To address this, some app developers are introducing companion toys that can be manipulated by little hands alongside the apps.

What we don't yet fully understand is how much value there is to the tactile element of touching interactive screens, something that requires a connection between the eyes, fingers and brain, and that passive viewing lacks. Does manipulating a digital object on screen enhance the learning process and make it easier to transfer knowledge into the physical world? And can understanding this mechanism help us develop better digital learning tools?

Regardless of our feelings towards tablets and smartphones, these devices are here to stay. So how do we get the most out of them? Thanks to some 100 years of research into how children learn, we can make educated guesses about what sort of interactions, in what sort of circumstances, are best.

Devices such as tablets and smartphones can make the most impact in lower-income households. In these households, people tend to have less access to developmental resources – such as music lessons, extra tuition or just extra hours of social interaction – and so spend more time with digital media. Provided the content is high-quality, tablets and smartphones can have a big impact.

For example, a study from Stanford University in the US found that, by 18 months, toddlers from disadvantaged families are already several months behind their more advantaged peers on language proficiency. With the right content and context, digital devices can help bridge the divide.

'It's a bit privileged and unrealistic to say no to technology,' says Levine. 'I worry that we are seeing people wagging their fingers at others because they do not have the privileges of time and resources that other families might have.

'There's no way we're going to improve the educational performance of young children without using technologies.'

Instead of banning devices, we should be demanding better apps built on solid research. For children aged between three and five, it's entirely possible that a well-designed app can help improve vocabulary and basic maths skills. 'My youngest is speech-delayed, and the videos he watches have definitely helped him learn new words,' says Lisa, a mother of four- and six-year-old sons who have been using mobile technology since they were 18 months old.

All of the paediatricians, child development and education specialists I spoke to agreed that, for children under 30 months, there is no substitute for human interaction. So why not develop apps that act as mediators between infant and caregiver? Bed-Time Math is one example. The app delivers engaging maths story problems for parents and their children. It is one of the few tools that have been shown to make kids smarter; children who used the app even just once a week for a year improved their maths by more than a control group did. The impact was particularly strong for children whose parents were anxious about maths.

With so much focus on what children are doing, it's easy for parents to forget about their own screen use. 'Tech is designed to really suck you in,' says Radesky, 'and digital products are there to promote maximal engagement. It makes it hard to disengage, and leads to a lot of bleed-over into the family routine.'

One approach that has been shown to help under-threes learn better is to build tools that use 'nudge technologies' geared at the parents. This could be text messages or emails that remind parents to sing or talk with their baby, to help both parents and child disengage from technology and apply learnings to the real world. Children's tablet maker LeapFrog does something similar with its LeapPad devices. Parents receive emails about what their child has learned from the touchscreen, along with ideas of how they could apply this new knowledge away from the screen.

'The extent to which parents are tied up with these devices in ways that disrupt the interactions with the child has potential for a far bigger impact,' says Heather Kirkorian, who heads up the Cognitive Development & Media Lab at the University of Wisconsin-Madison. 'If I'm on the floor with a child but checking my phone every five minutes, what message does that send?' How much parents play with and talk to their kids is a very powerful predictor of how the kids will develop, she adds.

Radesky has studied the use of mobile phones and tablets at mealtimes by giving mother–child pairs a food-testing exercise. She found that mothers who used devices during the exercise started 20 per cent fewer verbal and 39 per cent fewer nonverbal interactions with their children. During a separate observation of 55 caregivers eating with one child or more, she saw that phones became a source of tension in the family. Parents would be looking at their emails while the children would be making excited bids for their attention.

'You would see parents losing it and raising their voices because it's extremely irritating to be focusing on something and have a child escalate their requests for attention,' she explains, adding that some parents would do things like shove their child's hand away. Restricting the use of devices at critical family moments such as mealtimes and before bed can help reduce these frictions and encourage more face-to-face conversations.

Infants are wired to look at parents' faces to try to understand their world, and if those faces are blank and unresponsive – as they often are when absorbed in a device – it can be extremely disconcerting for the children. Radesky cites the 'still face experiment', which was devised by developmental psychologist Ed Tronick in the 1970s. In it, a mother is asked to interact with her child in a normal way before putting on a blank expression and not giving them any visual social feedback. As the video shows, the child becomes increasingly distressed as she tries to capture her mother's attention.

'Parents don't have to be exquisitely present at all times, but there needs to be a balance and parents need to be responsive and sensitive to a child's verbal or nonverbal expressions of an emotional need,' says Radesky.

§ § §

Although we are still in the early days of understanding the impact that mobile computers are having on young children, the key piece of advice from the child experts I spoke to was to make sure that device use is just one part of a rich diet of activities, particularly for under-threes, who seem to struggle to learn from screens.

Interactive, creative touchscreen experiences should be preferred over passive TV-like viewing. Parents should take educational claims from app developers with a hefty pinch of salt.

Where possible, a device should be used as a tool to enhance interactions with the child, whether that's as a launchpad for discussion ('What's the cow doing over there?' 'What sound does the duck make?') or as a way to inspire educational conversations that spill into the rest of the day, as appears to happen with BedTime Math.

Tronick's still face experiment did not involve screens, but a number of researchers have cited it as evidence that parents shouldn't be distracted by their smartphones when they are around their babies. This is true to an extent, but Tronick himself underplays its significance. 'It's all a bit exaggerated,' he says, adding that most children do plenty of activities every day that don't involve screens.

He is concerned that the worries about kids' use of screens is born out of an 'oppressive ideology that demands that parents should always be interacting with their child'.

'It's based on a somewhat fantasised, very white, very upper-middle-class ideology – tiger moms and helicopter parents – that says if you're failing to expose your child to 30,000 words you are neglecting them.' Tronick believes that just because a child isn't learning from the screen doesn't mean there's no value to it – particularly if it gives parents time to have a shower, do some housework or simply have a break from their child.

'Many parents, particularly low-income parents, are horrifically stressed and concerned they don't get the support they need and find parenting really lonely. Those are the big problems,' he says.

Parents can get a lot out of using their devices to speak to a friend or get some work out of the way. This can make them feel happier, which lets them be more available to their child the rest of the time. For Sandy, this is a relief to hear. 'Sometimes I'm at the end of my tether,' she says, adding that she shouldn't have to feel guilty about giving her child the iPad so she can have some 'me time'. With some parents, there's a lot of snobbery about screen use, she says.

'As a mum, I put my 18-month-old in front of an HBO baby poetry video,' says Radesky. 'It's cute and calm and I can wash the dishes or do something that's a reset for me. That's a benefit, but it's something parents need to be very honest about. The video is not educating my 18-month-old. It's a break for me as a parent.'

This story was first published on 7 June 2016
by Wellcome on mosaicscience.com

You can train a body into 'receiving' medicine

■ Jo Marchant

Marette Flies was 11 when her immune system turned against her. A cheerful student from Minneapolis, Minnesota, she had curly brown hair and a pale, moon-shaped face, and she loved playing trumpet in her high-school band. But in 1983, she was diagnosed with lupus, a condition in which the immune system destroys the body's healthy tissues.

It ran rampant, attacking her body on multiple fronts. She was given steroids to suppress her immune system; the drugs made her face swell up, and her hair fell out onto her pillow and into her food. But despite the treatment her condition worsened over the next two years, with inflamed kidneys, seizures and high blood pressure. She suffered frequent headaches and her whole body was in pain.

By 1985, antibodies were attacking a vital clotting factor in Marette's blood, causing her to bleed uncontrollably. It got so bad that her doctors considered giving her a hysterectomy, because they were worried that when her periods started she might bleed to death. She took drugs including barbiturates, antihypertensives, diuretics and steroids but her blood pressure kept rising. Then her heart started to fail, and her doctors reluctantly decided give her Cytoxan, an extremely toxic drug.

Cytoxan is very good at suppressing the immune system. But it causes vomiting, stomach aches, bruising, bleeding, and kidney and liver damage, as well as increased risk of infections and cancer, and at the time its use in humans was experimental. Karen Olness, a psychologist and paediatrician now at Case Western Reserve University in Ohio, was helping Marette to cope with the stress and pain of her condition, and she was concerned that if lupus didn't kill the teenager, this new drug might. Then Marette's mother showed Olness a scientific paper she had seen. It claimed to have slowed lupus in mice – but with just half the usual dose of Cytoxan.

The results were part of a well-known and seemingly mundane phenomenon that has been driving a quiet revolution in immunology. Its proponents hope that by cutting drug doses, it will not only minimise harmful side-effects but also slash billions from healthcare costs, transforming treatment for conditions such as autoimmune disorders and cancer. The secret? Teaching your body how to respond to a particular medicine, so that in future it can trigger the same change on its own.

§ § §

Ever eaten a favourite food that made you sick – prawns, say – and discovered that for weeks or months afterwards, you couldn't face eating it? This effect is called learned or conditioned taste aversion and it makes sense: avoiding foods that have poisoned us in the past protects us from getting ill again.

In 1975, a psychologist in New York was studying taste aversion in a group of rats and got an utterly mystifying result.

Robert Ader, working at the University of Rochester, gave his animals saccharin solution to drink. Rats usually love the sweet taste but for this experiment, Ader paired the drink with injections of Cytoxan, which made them feel sick. When he later gave the animals the sweetened water on its own they refused to drink it, just as he expected. So to find out how long the learned aversion would last, he force-fed this harmless drink to them using an eyedropper. But the rats didn't forget. Instead, one by one, they died.

Though Cytoxan is toxic, Ader's rats hadn't received anything close to a fatal dose. Instead, after a series of other experiments, Ader concluded that when the animals received saccharin and the drug together, they hadn't just associated the sweet taste with feeling sick, they'd also learned the immunosuppression. Eventually, they'd responded to the sweetened water just as they had to the drug. Even though the second phase of the experiment involved no drug at all, the doses of water Ader fed them suppressed their immune systems so dramatically that they succumbed to fatal infections. In other words, their bodies were reacting to something that wasn't really there, just because the circumstances made them expect it.

The phenomenon in which we learn to associate a contextual cue with a physiological response is well known. It's called conditioning and was discovered in the 1890s by the Russian physiologist Ivan Pavlov, who noticed that dogs learned to associate his presence with being fed, so that his arrival caused them to salivate even if he had no food. He showed that different signals – such as a buzzer or electric shock – could all be made to trigger the same automatic response.

Such learned associations are an important part of our daily lives. Cues prepare the body for important biological events such as eating or sex, and they trigger responses that have evolved to help us avoid – or flee from – danger. As well as inducing nausea, for example, exposure to a stimulus we associate with a previous allergic reaction (such as a grassy field or fluffy cat) can make us cough or sneeze even if no physical allergen is present, while previously scary situations (like a barking dog or enclosed space) can induce a state of fight-or-flight.

But Ader's result was revolutionary because it showed that learned associations don't only affect responses – such as nausea, heart rate and salivation – that scientists knew were regulated by the brain. His rats proved that these associations influence immune responses too, to the point at which a taste or smell can make the difference between life and death. The body's fight against disease, his experiment suggested, is guided by the brain.

In fact, a similar discovery had already been made in Russia. In the 1920s, researchers at the University of St Petersburg were following up on Pavlov's work, to see which other physiological responses could be conditioned.

Among them was the immunologist Sergey Metalnikov. Instead of suppressing the immune system, like Ader would, Metalnikov wanted to boost it. In one series of experiments, he repeatedly warmed guinea pigs' skin at the same time as giving them injections (small doses of bacteria, for example) that triggered an immune response. Then he gave them – and another group of guinea pigs that hadn't had this conditioning – a normally lethal dose of Vibrio cholerae bacteria, at the same time as warming their skin. The unconditioned animals died within 8 hours, Metalnikov reported, whereas the conditioned ones survived an average of 36 hours, and some of them recovered completely. Their response to a learned cue – the feeling of heat – appeared to have saved their lives.

Just like other learned associations, the phenomenon of conditioned immune responses makes evolutionary sense. Imagine that you encounter a pathogen – perhaps Salmonella bacteria in your prawn sandwich. As well as making you feel sick, this triggers a particular immune response. The next time you have a similar sandwich, your immune system doesn't have to wait for physical signs of bacterial invaders before mounting that response. Through conditioning, it can get one step ahead by triggering the same defence as soon as you taste or even smell the prawns.

The Russian studies weren't noticed in the West, however. And at first Ader's work was ignored too, largely because there was no known mechanism by which an animal could learn an immune response. The immune system and nervous system were thought to be completely independent, so Ader's theory that the two networks communicate was seen as crazy. Scientists were convinced that the immune system responds to physical signs of infection and injury without any help from the brain.

§ § §

'The time wasn't right for this new thinking.' Manfred Schedlowski, a medical psychologist at the University of Essen in

Germany, could be talking about Ader, but actually he's describing his own experiences in the mid-1990s, when he first set out to study conditioned immune responses for himself.

He was always interested in the links between mind and body, he tells me. At school, he enjoyed philosophy as much as physiology. His PhD investigated the effects of stress on the immune system in skydivers. As a researcher at Hannover Medical School, he turned his attention to conditioning, determined to transform the phenomenon described by Ader into a therapy that could be used to help patients.

He met obstacles straight away. On the hunt for other scientists to collaborate with, he knocked on the doors of the big immunologists. 'Some did not have time for me. Some listened to my story. One interrupted me after two to three minutes talking about the brain and the immune system. He said, 'Dr Schedlowski, if you want to do something like that, become an artist. That has nothing to do with science."

Undaunted, Schedlowski set about training rats to associate the taste of saccharin with the immunosuppressant effects of a drug similar to Cytoxan, called CsA. He found that their conditioned response to saccharin suppresses proliferation of white blood cells in their spleens, and cuts the production of two vital chemicals that the immune system uses for signalling (the cytokines IL-2 and IFN-?), just as the drug does.

Schedlowski wanted to know whether these conditioned responses could be medically useful. In particular, he thought they might be able to help with organ transplants, where a common risk is that the recipient's immune system will attack the foreign organ. To find out, Schedlowski transplanted second hearts into the abdomens of rats that had been conditioned with sweetened water and CsA, and then gave them daily doses of sweetened water alone. They tolerated the transplanted hearts for around 3 days longer than a control group (which had received sham conditioning with a placebo), and for as long as rats that had received no conditioning but got a short course of treatment with CsA after transplant. The conditioned response was as good as the actual drug.

A second trial, in which Schedlowski combined this conditioned response with very low doses of CsA, was even more dramatic. In unconditioned rats that got a low-dose course of CsA treatment, the transplanted hearts survived on average 8 days, the same as with no treatment. A full-dose course raised this to 11 days. But in rats that had the conditioned response plus low-dose CsA, the hearts survived on average 28 days, and more than 20 per cent of them lasted for several months, the full length of the experiment.

Schedlowski had feared that if learned associations weaken over time – a process known as extinction – then conditioned immune responses wouldn't be useful for patients on medication long-term. But by combining conditioning with a low drug dose, he says, 'we can interfere with this extinction'. Once the rats were trained, the combination of a sweet taste and just a tiny amount of the original drug protected the hearts. It was a stunning result that suggested Ader had been right about the power of conditioned responses, even in life-threatening situations such as organ transplants.

§ § §

A few years after Ader first published his findings, David Felten, then a neuroscientist at the Indiana University School of Medicine, found what the critics said was missing – proof that the immune system and nervous system were linked.

Felten was using a powerful microscope to track the paths of different nerves in the bodies of dissected mice. He was particularly interested in the autonomic nervous system, which controls bodily functions like heart rate, blood pressure and digestion. He found nerves connecting to blood vessels, for example, just as expected, but was flabbergasted to see them also running into immune organs such as the spleen and thymus. 'We were almost afraid to say anything,' he later told a reporter for PBS, in case he and his team had missed something and would 'look like a bunch of doofuses'.

But Felten's work checked out. It proved that there is a physical connection between nerves and immune cells. Felten moved to the University of Rochester to work with Ader and his colleague Nicholas Cohen, and the three are credited with founding the field known as psychoneuroimmunology, which is based on idea that the brain and immune system work together to protect us from illness. It's now known that communication runs in both directions, through hardwired nerves but also chemical messengers – cytokines and neurotransmitters – that speak to both the immune system and the brain.

Ader wondered if this new understanding could be harnessed to help patients. Conditioning had killed his rats, but could it treat disease, as in the Russian guinea pigs? Then he got a call about a girl who desperately needed his help.

§ § §

In a 1982 study, Ader had used conditioning to treat mice that had a lupus-like disease. He trained them to associate Cytoxan with saccharin solution, just as in his original experiment. After they learned the association, he kept giving the mice sweetened water along with half the usual drug dose for lupus. Compared to mice that received the same dose but weren't conditioned, their disease progressed more slowly and they lived longer. This was the paper that Marette's mother had seen.

Karen Olness telephoned Ader and asked: would his conditioning work on Marette? Could they train her immune system to respond to a lower drug dose than normal, sparing her from the worst of its toxicity?

Ader agreed to try.

The pair worked fast to design a conditioning regime for Marette. The first question was what taste to use. 'We had to choose something that was unique, that she hadn't experienced before,' says Olness. She considered vinegars, horehound, eucalyptus chips and various liqueurs before finally settling on a combination of rose perfume and cod liver oil.

The hospital's ethics board approved the trial in an emergency meeting and Marette's treatment started the next morning. She sipped the cod liver oil as Cytoxan flowed through an intravenous line into a vein in her right foot. Meanwhile Olness uncapped the rose perfume and waved it around the room.

They repeated this bizarre ritual once a month for the next three months. After that, Marette was exposed to cod liver oil and perfume every month, but received Cytoxan only every third month. By the end of the year, she had received just six doses of the drug instead of the usual twelve.

Marette responded just as her doctors would have hoped from the full drug amount. The clotting factor that her antibodies had been destroying reappeared, and her blood pressure returned to normal. After 15 months she stopped the cod liver oil and rose perfume but continued to imagine a rose, which she believed helped to calm her immune system. She graduated from high school and went to college, where she drove a sports car and played trumpet in the college band.

§ § §

At around nine o'clock every morning and evening, an alarm goes off on Barbara Nowak's mobile phone. When she hears it, the 46-year-old geologist sits down at the kitchen table of her home in Sprockhövel, northern Germany, and takes a powerful cocktail of immunosuppressant drugs. Their names – tacrolimus, Mowel, prednisolone – are now woven into the fabric of her life. But today there's a change to her daily routine. Before swallowing the pills, she pours herself a drink and downs it in one. It's sweet, bitter, neon green – and tastes strongly of lavender.

In 1988, when she was 19 and studying for high-school exams, Nowak lost her kidneys to lupus. She has spent many exhausting years since on dialysis, sitting 12 hours a week at her local clinic with huge needles in her arm – her flesh is still gouged with scars. Receiving a donated kidney transformed her health. 'It's another life,' she says. She has energy again and can travel – she now takes part in geocaching challenges across Europe with her pet beagle. But there's a downside. She's dependent on twice-daily medication to suppress the immune responses that would destroy her transplant.

The drugs keep her kidney working but have side-effects, from tremors and nerve pain to gum disease and growth of facial hair. Nowak has been lucky enough so far to avoid the worst of these: although one drug started to destroy her red blood cells, since switching to an alternative she is dealing with her medication well. But she knows she is at increased risk of life-threatening infections, heart failure and cancer. And the drugs slowly poison the very organ she's trying to save.

So Nowak is drinking this gaudy concoction as part of a pioneering trial at the nearby University of Essen. The 'famous green drink' – as Schedlowski's students like to call it – is an updated version of Marette's rose and cod liver oil, invented to test conditioned responses in people. Like Ader, Schedlowski wanted something strange and unforgettable that stimulates several senses at once. He hit on strawberry milk mixed with green food colouring and essential oil. Its bright colour and overwhelming lavender flavour creates a bewildering mix of sensory cues, like drinking a violent, bittersweet battle between green and purple.

So far, Schedlowski has shown that after being associated with CsA, the drink reliably induces immunosuppression in healthy volunteers, creating on average 60–80 per cent of the effect of the drug. And just as in the rats, combining the conditioned response with a low drug dose prevents the learned association from fading. But will it work in patients?

I'm with Nowak on the trial's last day. She is small but looks strong, and her tanned face is etched with smile lines. She says she was already familiar with the power of conditioning after using clicker training with her beagle, Ivy, and loved the idea of trying it on herself. 'I thought it was so funny,' she says.

She removes her fleece to reveal a T-shirt with a stethoscope printed on it, then a research assistant hands her a 50-ml centrifuge tube, full to the brim with the green lavender milk. It's the brightest thing in the room. 'Danke schön!' she says. She gulps it fast, makes a face, and reaches into her rucksack for a sweet to take away the taste.

Schedlowski is running this trial with Oliver Witzke, a nephrologist at Essen's University Hospital. For Witzke, the dangers of high drug doses are agonisingly real. He spends his career prescribing powerful immunosuppressants including CsA to kidney transplant recipients like Nowak. 'Quite a lot of our patients die prematurely,' he says. 'Not because the transplant fails... It's due to the drugs that I prescribe every day.'

In every transplant patient he cares for, getting drug doses right is a delicate balancing act. Get the dose too low, and the patient will reject the kidney. But get it too high, and you'll destroy the kidney or kill the patient. 'We lose about 10 per cent of transplants in the first year,' says Witzke. Half of those patients go back on dialysis, the other half die. After that, the rate of decline slows, but some kidneys are still lost each year, and patients are at an increased risk of death due to drug complications.

One of the most damaging side-effects is nephrotoxicity: the drugs directly destroy kidney cells. The average life of a transplanted kidney is eight to ten years, says Witzke, and often when a kidney fails it isn't clear whether the underlying cause is rejection or toxicity. 'The dream for every transplant person is not to feed the patient from the first hour with a drug that's toxic for the transplant.'

The search for immunosuppressants that aren't nephrotoxic hasn't been successful so far, but Witzke hopes that using conditioning to reduce doses will keep his patients alive longer. When he first heard about the concept, 'I thought it was rubbish,' he admits. 'As a doctor, I believe in pharmaceutics and drugs.' But Schedlowski's experiments convinced him that the effect is not just a psychological trick. 'It has a biochemical basis,' he says.

At this stage it's too risky for Nowak and her fellow trial participants to reduce their drug doses, so the first step is to see if conditioning can suppress their immune systems over and above the effect of their normal pills. Some of the participants are taking CsA, but Nowak is on tacrolimus. In the learning phase of the study, she drank the lavender milk alongside her drugs, morning and evening, for three days. Then, after a two-day break, came the 'evocation' phase, using the green drink to try to amplify the effect of her medication. She again downed the drink with her drugs, but this time, she drank it two extra times during the day, along with a placebo pill.

A pilot trial carried out in 2013 was promising: in all four patients, adding the green drink suppressed immune-cell proliferation and levels of the signalling molecule IL-2 by 20–40 per cent more than drugs alone. Now Nowak is part of a larger study of around 20 patients. If that works too, the next step will be to test whether this conditioned response can maintain immunosuppression while drug doses start to be reduced.

The hope is that this will reduce unwanted side-effects. Some problems, like infection risk, are likely to be an inevitable consequence of suppressing the immune system, whether that's achieved using drugs or lavender milk. Others, like nausea, will perhaps be conditioned along with the immunosuppression. But Witzke argues that side-effects caused directly by drug toxicity – including kidney damage and increased cancer risk – are unlikely to accompany conditioned immune responses. It won't be possible to lose the drugs completely, he says, but he hopes that even reducing doses by 20–30 per cent would improve quality of life while prolonging the survival of transplanted kidneys to perhaps 12 or 15 years.

Nowak isn't convinced by the drink itself. 'It's awful!' she says. The taste got worse the more she drank, she explains, and she didn't like carrying the odorous liquid around with her all day. An odd-tasting candy might be more practical and palatable, she suggests. But she's right behind the principle of the trial, describing anything that might preserve her kidney as 'very important'.

At 46, she is already on her third transplanted kidney. The first failed after a week, the second after 13 years – possibly because of drug toxicity – and her doctors say that after five years, her current kidney is ageing more quickly than expected. 'It would be better if this one lasts longer,' she says bluntly. If it fails, she faces more years on risky, exhausting dialysis – average life expectancy on dialysis is just five to ten years – and the agonising wait for another donor.

§ § §

Besides helping with organ transplants, there's a plethora of uses that conditioning might have, by reducing harmful side-effects or simply making treatment more cost-effective for patients and governments that can't afford constant full doses of the most expensive drugs. Other possibilities include allergies and autoimmune conditions.

For example, Ader carried out a small study in 1996 that paired Cytoxan with aniseed-flavoured syrup in ten people who had the autoimmune condition multiple sclerosis. When later given the syrup alongside a placebo pill, eight of them responded with immunosuppression similar to that produced by the active drug. In another study, published shortly before he died in 2011, Ader reported that quarter- or half-doses of corticosteroid ointment plus conditioned responses could control psoriasis just as well as a full drug dose.

Schedlowski aims to test that psoriasis result in a pilot study planned for spring 2016. He has already shown that after conditioning with the antihistamine drug desloratadine, the green drink reduces immune responses and symptoms in people allergic to dust mites. And he is collaborating with Rainer Straub, an immunologist at the University of Regensburg, Germany, to study conditioned immune responses in rats with a model of arthritis. The results are not yet published, but Straub says that so far, conditioned responses plus a low drug dose appears to suppress the inflammatory response 'even better' than full-dose drug alone.

Animal studies hint that the approach might also be useful in the treatment of some cancers. In the 1980s and 90s, researchers at the University of Alabama, Birmingham, trained mice to associate the taste of camphor with a drug that activates natural killer cells – white blood cells that attack tumours. Then they transplanted aggressive tumours into the mice. After the transplant, mice given doses of camphor survived longer than those treated with immunotherapy, and in one experiment, two mice defeated their cancer altogether, despite receiving no active drug. Schedlowski is following up on these results too, and so far has shown that the effects of the anti-tumour drug rapamycin, which stops immune cells from dividing, can be conditioned in rats.

Key questions include pinning down the precise mechanism of conditioned immune responses, and working out why some individuals respond more strongly to conditioning than others. 'Some people respond very nicely, but others don't respond at all or they show only a minor response,' says Schedlowski. So far, he has discovered that the effect is mediated by the sympathetic nervous system, which drives our response to stress and is part of the network that Felten discovered linking the brain and immune system. In experiments where Schedlowski cut the nerve running to rats' spleens, the conditioned response was completely blocked. Intriguingly, he has also found that people with high levels of anxiety, and of the stress hormone noradrenaline, respond better to conditioning, possibly because they have a more active sympathetic nervous system.

Another important area for future research is looking at which physiological responses – not just immune responses but among other systems too – can be conditioned. For example, Schedlowski hasn't been able to condition the effects of corticotropin-releasing hormone, which is involved in the stress response. On the other hand, learned associations are known to be strong in pain and psychiatric disorders such as depression. It's one reason why placebos are so effective in these conditions: our bodies learn the appropriate physiological response to pills we take and will subsequently repeat it, for example releasing pain-killing endorphins, even if a pill contains no active drug.

Years of research are required before conditioning regimes for cancer or transplant patients reach the clinic, but Schedlowski says the principle could be used much sooner to reduce drug doses for non-life-threatening conditions such as asthma or arthritis. Paul Enck, a medical psychologist at the University of Tübingen, Germany, agrees. He suggests a method that he calls 'placebo-controlled dose reduction'. For example: when someone is prescribed a suitable drug, after two or three weeks of taking it regularly they could switch to a pack in which their pills are interspersed with identical placebos.

In a 2010 trial, children with attention deficit hyperactivity disorder (ADHD) were asked to take a distinctive green-and-white placebo pill alongside their drugs. The children knew these pills were placebos. But those who went through this conditioning process later did just as well on the placebo plus half their normal drug dose as another group did on the full drug dose – and significantly better than children who received a half-dose without conditioning. If used widely, advocates say, substituting some of the drugs we take for placebos could save billions of dollars in healthcare costs. In the US, for example, drugs for ADHD alone cost more than $5.3 billion a year.

But the idea is not widely accepted. That's perhaps partly because the prospect of reducing drug doses isn't attractive for drug companies, which drive most research and development into new therapies. 'They don't like the story very much,' says Schedlowski. 'They see their drug and their marketing jeopardised.' A wider problem is that for most doctors and scientists, the concept of treatments with no pharmaceutical component just makes no sense.

This scepticism is familiar to Adrian Sandler, a paediatrician at the Olson Huff Center for Child Development in Asheville, North Carolina, who carried out the 2010 ADHD trial. He says he'd love to run more trials to see how reducing drug doses might help with ADHD and other disorders such as autism, but his applications for funding have been rejected. 'I think it's a highly unusual kind of study,' he says. 'The idea of using placebos in open-label to treat a condition is innovative, it turns things upside down. Some reviewers may find that hard to accept.'

When Ader and Karen Olness published Marette's case, they were careful to say there was no proof that she wouldn't have done just as well without the conditioning. Schedlowski has since built a strong case, however, that immune responses can be conditioned in humans with wide-ranging potential benefits. Marette, despite the initial success of her Cytoxan treatment, didn't live to see it. She died aged 22, in February 1995. According to Olness, the toxic drugs she took earlier in life had irreparably damaged her heart.

Twenty years on, is Nowak likely to see benefits in her lifetime, or is resistance to this unconventional approach simply too strong? Brian Ferguson, who studies the innate immune system at the University of Cambridge, offers some hope. He thinks we're on the verge of a 'snowball' of research interest in brain–immune connections, driven in part by growing awareness of the importance of inflammation in neurodegenerative disease. That's helping to break down barriers between neuroscience and immunology, he says, and might ultimately help acceptance of behavioural studies too.

Meanwhile Schedlowski is steadfastly optimistic that the benefits of conditioning are too great to ignore. 'Ten years ago, nobody believed us,' he says. 'Now, journals are much more open-minded to this kind of approach.' He believes that within a decade or two we'll see a revolution in which learning regimes will become a routine component of drug treatment for a wide range of conditions. Drug companies might not see the advantages now, but in future, he argues, they could use the reduced side-effects of lower doses as a selling point.

For now, though, there's a long way to go before the potential for conditioned immune responses is widely accepted, let alone used in the clinic. It's hard enough for people to entertain the idea of using placebos to treat pain, or psychiatric disorders, and using them to influence immune responses sounds even crazier.

Brain–immune interactions are a 'blind spot' for immunologists, admits Ferguson, with funding and interest for this type of work practically non-existent. Researchers are 'vaguely' aware that the two systems communicate, he says, 'but there's this traditionality whereby people describe the immune system as everything going on from the neck downwards, and the central nervous system is everything from the neck upwards, and the two things haven't been linked very much.'

Pavlov won a Nobel Prize for showing that the digestive system, previously thought to function independently, is in fact tightly controlled by the brain. Despite showing that the same is true for the immune system, Ader and Felten are barely known, even among immunologists. Schedlowski, supported by the DFG (the German Research Foundation), leads one of the only teams researching conditioned immune responses. 'I like to say we're the best in the world,' he jokes. 'Because there is nobody else!'

This story was first published on 9 February 2016
by Wellcome on mosaicscience.com

Charting the phenomenon of deep grief

■ Andrea Volpe

After Stephanie Muldberg's 13-year-old son Eric died of Ewing's sarcoma in 2004, she was lost in a sea of grief. Her days were long, unstructured, monotonous. She barely left her New Jersey home. When she did leave, she planned her routes carefully to avoid driving past the hospital, just a few miles away, where Eric had been treated during the 16 months of his illness, or the fields where he had played baseball. Grocery shopping was a minefield, because it was painful to contemplate buying Eric's favourite foods without him. To enjoy anything when he could not felt wrong. And Muldberg never thought she would be able to return to the temple where he had celebrated his bar mitzvah – and where his funeral was held.

Looking back, she describes herself as not knowing how to grieve after Eric died. 'I didn't know what to do, how to act in front of people – what I needed to do privately, who I could reach out to. I was fearful of making people more emotional, too emotional, and having to comfort them,' she tells me, by Skype. 'I didn't know how to talk about what I was thinking.' Muldberg's long dark hair is pulled back and she's wearing a white T-shirt. One of the things she says is that she thought if she stopped grieving, her memories of Eric would fade, and she'd lose her connection to her son for ever.

The passage of time often seems the only remedy for grief, but time didn't help Muldberg. In the years following Eric's death, she says, she felt consumed by grief. Then a family physician heard a talk by Columbia University psychiatrist Katherine Shear about treating chronic and unremitting grief and thought Shear might be able to help her.

Four years after Eric died, Muldberg arrived at the New York State Psychiatric Institute in Manhattan, for her first meeting with Shear. She answered Shear's questions with as few words as possible. It was as if she were barely present in the small, windowless room. Her face was drawn and clouded; she sat crumpled in her chair, arms crossed tightly around her, as if the weight of her loss made it impossible to sit up straight. It felt to her as if Eric had died just the day before. Shear diagnosed Muldberg with complicated grief, the unusually intense and persistent form of grief she has been researching and treating for almost 20 years.

Grief, by definition, is the deep, wrenching sorrow of loss. The initial intense anguish, what Shear calls acute grief, usually abates with time. Shear says that complicated grief is more chronic and more emotionally intense than more typical courses through grief, and it stays at acute levels for longer. Women are more vulnerable to complicated grief than men. It often follows particularly difficult losses that test a person's emotional and social resources, and where the mourner was deeply attached to the person they are grieving. Researchers estimate complicated grief affects approximately 2 to 3 per cent of the population worldwide. It affects 10 to 20 per cent of people after the death of a spouse or romantic partner, or when the death of a loved one is sudden or violent, and it is even more common among parents who have lost a child. Clinicians are just beginning to acknowledge how debilitating this form of grief can be. But it can be treated.

I first learned about complicated grief while riding the subway in Boston, where I read an advertisement recruiting participants for a study at the Massachusetts General Hospital, which I later discovered was related to Shear's research. By then, I'd been a widow for about a decade. I was 33 when my husband died and it was fast – just six weeks from when he was diagnosed with pancreatic cancer. My grief had a different kind of complication: I was pregnant, and our son was born seven months after his father's death. By the time I read that subway ad, he was in elementary school, and I was holding my own. I gradually went back to work. Single parenting was overwhelming, but it kept me focused on what was right in front of me. Having a young child is filled with small pleasures and motherhood enlarged my sense of community. I fell in love again. But it still felt like I walked with a limp, and that limp was grief.

Often, I felt that the course of my grief – as it slowed or accelerated – wasn't within my control. Sometimes I'd buckle, and wait it out. Sometimes I'd push back. Somehow, I knew it was going to take as long as it took. There wasn't anything to do about it except live. Freud, writing in Mourning and Melancholia, one of the first psychological essays on grief, saw it this way, too: 'Although mourning involves grave departures from the normal attitude of life, it never occurs to us to regard it as a pathological condition and to refer it to a medical treatment. We rely on its being overcome after a certain lapse of time, and we look upon any interference with it as useless or even harmful.' That's how it went for me.

I'd be the first to say that my path through grief has been intellectual. I've spent years contemplating what grief is. That subway ad made me wonder: Was my grief a disease? To be diagnosed with an illness is to seek – or wish for – a cure. But conceiving of grief as a disease with a cure raises questions about what is normal – and abnormal – about an experience that is universal. Is grief a condition that modern psychology, with its list of symptoms and disorders and an ample medicine cabinet, should treat, as if it were an illness rather than an essential part of being human?

A little more than a year ago I began sitting in on clinical training workshops at Columbia's Center for Complicated Grief, which Shear directs. The first workshop was both a challenge and a relief. It was strangely comforting to be in the company of so many people – grief counsellors, social workers and therapists – who spent their time thinking about what it meant to grieve. It would be almost another year until I called Stephanie Muldberg to see if she'd be willing to talk at length about what her treatment was like.

Sometimes I can feel in our conversations how deliberately she chooses her words. She is, she tells me, a very private person. At times her desire to talk about her experience of complicated grief feels in tension with her natural inclination to be more self-contained. 'I think the problem is people don't talk about grief, and I want to normalise the fact that people can talk about it, and make it easier, and not so taboo,' she tells me.

§ § §

For something so fundamental to being human, there's still a great deal we don't know about the grieving process. It wasn't until the 20th century that psychologists and psychiatrists claimed expertise over our emotions, including grief. The conventional wisdom about grieving is that it's something to be worked through in a series of stages. Lingering on any stage too long, or not completing them within a certain window of time, might be dysfunctional. Clinicians disagree about how long is too long to grieve, about whether the grieving person should wait for her grief to shift on its own or do something to initiate that process, and about what to do, and what it means, if grief is slow or stalled.

The idea of grief as something we need to actively work through started with Freud. John Bean, a psychoanalyst who has trained extensively with Shear and worked with her to treat patients in her research studies, explains to me that because Freud believed we have a limited supply of psychological energy, he viewed the central emotional 'task of grieving' to be separating ourselves emotionally from the person who died so that we can regain that energy and direct it elsewhere. Freud thought this would take time and effort and it would hurt. His theory of 'grief work' persists, often in tandem with newer theories of grief.

If grief is work, then Elisabeth Kübler-Ross provided the directions for how to do it. Kübler-Ross first proposed the five-stage model in 1969 as a way to understand the psychology of the dying, and it quickly became a popular way to understand bereavement. Today, those stages – denial, anger, bargaining, depression and, finally, acceptance – are practically folklore.

But it turns out grief doesn't work this way. In the past several decades, more rigorous empirical research in psychology has challenged the most widely held myths about loss and grief.

When George Bonanno, professor of clinical psychology at Columbia University's Teachers College, researched the paths people take through grief, he discovered there's more variation to how we grieve than psychologists thought. His office, in a massive gothic brick building in New York City's Morningside Heights, is crammed with books and lined with Chinese sculptures. On a rainy afternoon he outlines the three common paths he identified. Some people, whom he terms 'resilient', begin to rebound from loss in a matter of weeks. Others adapt more gradually, following a 'recovery' path. The intensity of those first days, weeks and months of mourning subsides. They 'slowly pick up the pieces and begin putting their lives back together', typically a year or two after losing someone close to them. People with complicated grief, like Muldberg, struggle to recover. Their grief becomes what Bonanno calls 'chronic', staying at a high level of intensity for years.

One school of thought that has influenced Shear is called the dual-process model: grief is stressful, so we alternate between confronting the emotional pain of our loss and setting it aside. Even grieving people, research has shown, have moments of positive emotion in their lives. Hope returns gradually. If the stage model maps a single, clear path through grief, then the dual-process model could be seen as a charting a wave pattern through grief.

It's now an axiom of grief counselling that there's no one right way to grieve. That seems like a good thing, but it's also a problem. If everyone grieves differently, and there's no single theory of how grief works, then who's to say that someone like Muldberg isn't making her way through grief in her own way, on her own clock? Even though it was clear to her and to those around her that, four years after her son's death, she was still suffering, bereavement researchers don't agree about how to explain why her grief was so prolonged – or what to do about it.

§ § §

Shear, who is in her early 70s, is the warmest shrink you'll ever meet. Everything about her conveys equanimity, especially the way she can sit with the stories of patients whose grief is unrelenting.

It wasn't always that way. 'At the beginning,' she tells me, she was 'afraid to sit in the room with someone who was really intensely grieving because I was still a little bit uneasy with death and dying, but also because it makes you feel so helpless – because you feel like there's nothing you can do'. The grieving person, she says, 'feels like the only thing that's going to help' is bringing back the person they are grieving – 'and you agree'.

'Grief is not one thing,' Shear says. 'When it's new, it crowds out everything else, including even people and things that are actually very important to us. It stomps out our sense of ourselves, too, and our feelings of competence. We think of grief as the great disconnector, but over time, it usually settles down and finds its own place in our lives. It lets us live in a meaningful way again. It lets us have some happiness again.'

Two weeks later, I'm jammed into a hard plastic desk in an overheated university classroom listening as Shear, who is professor of psychiatry at Columbia's School of Social Work, explains the underlying principle of her work, which is that 'grief is a form of love'.

She quotes me C S Lewis's *A Grief Observed* to explain what she means: 'Bereavement is an integral and universal part of our experience of love. It is not the truncation of the process but one of its phases; not the interruption of the dance, but the next figure.' This is called an attachment approach to grief. It's shared by many grief researchers and counsellors, and it can be traced back to the British psychiatrist John Bowlby. Attachment is what gives our lives security and meaning. When an attachment is severed by death, Shear says, grief is the response to the lost attachment. Peel back the psychological theory, and what you'll find is something that anyone who has experienced grief knows intuitively: 'Nature is so exact, it hurts exactly as much as it is worth, so in a way one relishes the pain, I think. If it didn't matter, it wouldn't matter,' writes the novelist Julian Barnes in Levels of Life, his extended essay on grief following the death of his wife.

Shear explains that it's our close bonds to those dearest to us that also help us want to care for other people and confidently explore the world. These attachments are woven into our neuro-biology. The longing and yearning of acute grief, and the feeling of unreality that comes with it, she says, are symptoms of just how much grief short-circuits our bio-behavioural wiring.

Shear agrees with Bonanno that over time most grieving people integrate their loss into their lives. But people with chronic grief face some complicating factor. Complicated grievers tend to be women. They are often excellent carers but not so good at taking care of themselves or accepting help. Often, their emotional re-serves of self-compassion and self-motivation have been drained. Shear says that 'we don't grieve well alone', but frequently people with complicated grief become isolated because their grief has remained at high levels for so long; the people around them may feel that they 'should have gotten over it by now'.

Shear believes that adapting to grief and loss is 'a normal, nat-ural process', she says. 'We're not talking about grief itself being abnormal. We're talking about an impedance in some problem of adaptation.' Think of it this way: her therapy jump-starts a stalled process, the way a defibrillator restarts a stopped heart.

§ § §

Shear's office, with its striped beige wallpaper and mahogany furniture, is so spotless it would feel like a hotel room it if weren't for the picture of her grandson as a chub-cheeked toddler on her panoramic Apple monitor. It's a sticky day in July, and she's telling me how she came to study and treat grief.

In the 1990s, Shear was researching anxiety and panic disorders at the Western Pennsylvania Psychiatric Institute and Clinic when she became involved with research on depression and anxiety in elderly people. One of the common triggers for depression in the elderly is the death of a spouse, and the team she was working with identified a cluster of symptoms in depressed patients that weren't depression. They expressed deep yearning, were often driven to distraction by thoughts of their deceased spouse, and had great difficulty accepting death, to the point that persistent, acute grief became a risk to their physical and mental health.

To differentiate grief-related symptoms from depression and anxiety, Shear worked with a research team that included psychiatric epidemiologist Holly Prigerson. It was Prigerson who, in 1995, had published a questionnaire that identified complicated grief as a specific syndrome and could accurately assess its symptoms. Shear has relied on it as a diagnostic and assessment tool in her research ever since. Shear and her colleagues also used it to design a new treatment, complicated grief therapy. Prigerson, who now holds an endowed professorship at Weill

Cornell Medicine in New York City, and directs Cornell's Center for Research on End-of-Life Care, continues to work on the epidemiology of prolonged grief.

§ § §

In their first meeting, Shear asked Stephanie Muldberg to keep a daily grief diary, recording and rating her highest and lowest levels of grief. Muldberg kept this diary for the duration of the therapy. Every day for almost half a year she was paying such close attention to her grief that it became inscribed in her daily life. Not that her grief wasn't already a pronounced everyday presence, but now, with Shear's help, she was facing it head-on rather than avoiding it. The diary was one of several techniques Shear used to help Muldberg look her grief in the eye.

Muldberg says that the grief diary helped her pay attention to herself in a way she hadn't been able to do in the four years after Eric's death. Using the diary, she began to see that she had some happy moments interspersed with some low times of grief. 'There were always going to be hard times during the day for me, but I wasn't only focusing on the hard times, I was starting to learn how to move forward.'

Complicated grief therapy (CGT) takes place over 16 sessions, structured, Shear says, by techniques adapted from approaches used to treat anxiety disorders, including cognitive behavioural therapy, a well-researched approach to psychotherapy, and exposure therapy, used to treat avoidance and fear in anxiety disorders. The structure itself is part of the therapy, she says, because structure is reassuring to people who are feeling intense emotions.

Shear has been testing CGT since the mid 1990s. In 2001, she and her colleagues published a small pilot study that showed

promising results. Since then, they have published several randomised controlled studies supported by the National Institute
of Mental Health, demonstrating that CGT helps patients who
have complicated grief to reduce their symptoms better than
conventional supportive grief-focused psychotherapy. Shear
is a pioneer, but she's not an outlier. Currently a group therapy version of CGT is being studied at the University of Utah.
Researchers in the Netherlands and Germany are also exploring variations on cognitive behavioural therapy and exposure
therapy to treat traumatic and prolonged grief. And a recent
study in Wales confirms one of Shear's main findings, which is
that the techniques in her treatment are more effective together
than separately.

§ § §

A few sessions into her treatment, Shear asked Muldberg to do
something she had never done, which was to tell the story of the
day Eric died. It's a technique Shear adapted from prolonged
exposure therapy that she calls 'imaginal revisiting'. At first,
Muldberg says, she was apprehensive because she wasn't sure if
she could remember what had happened. Over the course of three
weekly sessions, Muldberg told the story of Eric's death, rating
her levels of emotional distress as she did. The purpose of this
technique is to 'help people connect with the reality of the death
in the presence of a supportive person who is bearing witness to
it,' Shear explains. 'We want to keep grief centre stage,' she says.
'If you do let yourself go there, paradoxically your mind finds a
way to face that reality and to reflect on it.'

Then, as with the grief diary, Muldberg had 'homework': listening to a tape of herself telling the story every day between

sessions. At first, this was distressing, but she gradually learned how to manage her emotions, recognising, she tells me, that she wasn't going to forget Eric. The intensity of her feelings began to lessen, so that by about halfway through the therapy she began to feel better.

Muldberg admits she was sometimes sceptical of what Shear was asking her to do, and she says sometimes she pushed back. Part of CGT includes psychoeducation, in which the therapist explains to the patient the premise and purpose of the therapy. Shear's explanations, Muldberg says, helped her understand that 'there was a reason I was feeling this way'. She describes Shear's approach as 'I don't want to push you but we're going to figure out ways that you can accomplish these things, feel good about them, and do them.'

A few weeks after Muldberg started revisiting the story of Eric's death, she worked with Shear to make a list of the places and activities she had been avoiding since he died, and gradually started trying to face them. Shear calls this 'situational revisiting', a form of prolonged exposure therapy. 'We do this to provide people with an opportunity to confront the reality of the loss and actually understand its consequences, because being there without the person is going to be different than being there with the person. We want people to start to reflect on that,' she tells me.

For Muldberg, many of the things she had avoided were the everyday parts of being a mother, such as going to the grocery store, but she says, 'I didn't realise how much harder avoidance was than doing some of these things.' Together with Shear, she broke down tasks, such as driving past the baseball field where Eric had played, into smaller steps until she could do them again.

§§§

Sitting in that classroom listening to Shear explain these exercises makes my chest tighten until my heart aches. I can't imagine doing them myself, let alone how anyone with complicated grief could withstand them. It seems like a wrenching exercise in repeatedly tearing a scab off a wound.

When I ask Shear about this she acknowledges that her approaches are counter-intuitive because they 'ask people to go toward their grief'. She tells me it's by explicitly detailing and describing their grief that people with complicated grief become unstuck, as they learn to shift back and forth between the pain of grief and restoring their lives. Shear is more interested in having patients engage with the therapy techniques than she is with getting them to reach a certain point. To her workshop audience, she puts it this way: 'We do not try to lower grief intensity. I'm just trying to turn the Titanic one degree.'

In one of my conversations with Muldberg, I remark that CGT seems counter-intuitive, almost confrontational, and that these exercises seem extremely emotionally demanding. She is quick to correct me. Therapy was challenging, she says, but it came as a great relief to finally feel understood and have the support to face Eric's death. 'When I started to do things, I started to feel better,' she tells me.

For Shear, 'feeling better' is a sign that our natural adaptive abilities are kicking in, allowing a person who is suffering from complicated grief to begin the emotional learning process that ultimately helps grief subside. This also creates an opening for the person to begin to reimagine their life after a devastating loss.

At the same time that Shear was helping Muldberg come to terms with the reality of Eric's death, she was also helping her begin to envision the future. Part of losing someone very close,

Shear says, is that we lose our sense of identity. Part of grieving is regaining it.

In another CGT exercise, the therapist asks a scripted question: 'If someone could wave a magic wand and your grief was at a manageable level, what would you want for yourself? What would you be doing?' Someone with complicated grief can't imagine a future without the person they've lost, or without the unrelenting, intense grief that's taken up residence in their life. It's a future-oriented question for someone who has lost sight of the future. Just asking the question, Shear says, can activate our innate exploratory system and spark hope.

One way to think of the therapist's role in CGT is that she's teaching her patient what grief is. 'Loss is a learning process. The problem is, it's unwanted information,' says therapist Bonnie Gorscak, one of Shear's long-time collaborators and a clinical supervisor at the Center for Complicated Grief. Learning from loss, Gorscak says, means being able to 'stand in a different place and look at grief', to approach the pain it causes, experience it, and have some respite from it. It's a counter-intuitive approach for therapists, too. Sitting with someone with complicated grief, Gorscak says, 'is some of the worst pain I've ever sat with'.

§ § §

CGT is challenging, but it works. Still, Shear's therapy has sparked controversy, starting with the very idea that there is a form of grief so severe and debilitating that it meets the definition of a mental illness.

In recent years, Shear and a group of colleagues have advocated for a grief disorder to be included in the Diagnostic and Statistical Manual (DSM), psychology's diagnostic bible, because they believe complicated grief is a clear-cut, diagnosable syndrome, separate from depression, anxiety or post-traumatic stress disorder. (Shear and Prigerson, once collaborators, now disagree about the best way to diagnose complicated grief, but they agree it should be viewed as a mental disorder.) Without sanction by a DSM diagnosis, psychotherapy in the US is not covered by health insurance. Without insurance reimbursement, CGT is out of most people's reach. In 2013 the DSM-5 listed Persistent Complex Bereavement Disorder as a 'condition for further study', calling for more research on the issue.

The major issue therapists have with complicated grief is that they believe it pathologises a fundamental human experience. Leeat Granek, a health psychologist at Israel's Ben-Gurion University, is concerned that including a grief disorder in the DSM could narrow the spectrum of acceptable ways to grieve and create a narrative that would distort the ways people understand their own grief. She believes that this would lead to 'a lot of shame and embarrassment for the mourner because the expectations around grief are no longer realistic'.

Donna Schuurman, senior director of advocacy and training at Portland's Dougy Center, which supports grieving children and families, questions the idea of a grief disorder. She rejects the use of terms such as 'complicated', 'debilitating' or 'persistent' to describe grief reactions and as the basis for constructing a diagnosable syndrome. Schuurman agrees that 'grieving people may have chronic issues or chronic problems related to what has happened after someone dies', but says that 'often those issues were already there before the death', and that 'chronic issues ought not to be framed as mental disorders of grief'.

'Medicalizing or pathologizing the experience of someone who is having difficulty after a death does not do justice to the full social and cultural context in which he or she is grieving,' she writes. 'Grief is not a medical disease, it is a human response to loss. Many people who are experiencing severe challenges after a loss are doing so because the social expectations around them are not supporting them.'

Instead of labelling complications of grief as symptoms that define a disorder, Schuurman says she would focus on the experiences and behaviours that were contributing to any 'serious challenges' a grieving person was facing. 'We can label it depression, drug or alcohol abuse, etc., as any good therapist should do,' and 'try to look at underlying issues, and not just symptoms, to be of help,' she explains. Good professional help, she believes, 'could take a variety of forms and theoretical backgrounds'.

New scientific research on grief, Shear's among it, is challenging some of the foundational premises of grief counselling as it has been practised, often in community settings. As George Bonanno discovered, there are several common trajectories through grief, meaning that there are some commonalities among grieving people as they adapt to loss. Still, Shear says, 'each experience of grief is unique, just as each love experience is unique'. CGT, she says, 'helps people find their pathway to adapting to loss'.

§§§

One way to answer the question of whether or not grief is a disease is to ask if the treatment provides a cure. Stephanie Muldberg describes her grief as 'a wound that wasn't healing', but CGT isn't a cure the way antibiotics cure an infection. Grief doesn't end, it just changes form. Muldberg says CGT taught her how to live with grief as part of her life. She still carries her grief for Eric with her, but she is also back in the world. She travels with her husband and daughter. She volunteers for the Valerie Fund, an organisation that supports families of children with cancer and blood disorders, and that helped Eric and their family when he was sick.

§ § §

I ask Shear when her fear of sitting with intensely grieving people had subsided. 'Well,' she says, 'there's this entire field of study called terror management.' I was expecting her to tell me about her feelings but she answers by telling me how research explicated them – exactly what she's done in designing a therapy for complicated grief. I look up terror management: it's the theory that in order to deal with the fear of our own mortality, we find ways to find meaning and value in our lives – like helping people. In that sense, what Shear has done with CGT is to create a form of evidence-based compassion. It's compensation, perhaps, for the existential helplessness of the therapist, but it also compensates for many of our communal failures helping people grieve. We are too busy, too secular, too scared to deal with grief. It's hard for Western culture – American culture in particular – to sit with something that can't be fixed.

§ § §

The more I thought back over my conversations with Stephanie Muldberg, the more I thought about how her therapy with Shear helped her put Eric's death in context of her life story. The idea that a story needs a beginning, middle, and end goes back to Aristotle. People with complicated grief can't see the arc of their own stories. They can't get to what classic plot theory calls denouement – resolution. Most of us, when faced with a loss, find a way of putting what happened into the form of a story: this is what happened, this is who I was, this is what the person who died meant to me, and this is who I am now. But people who have complicated grief can't do this.

Grief is a problem of narrative. A story, in order to be told, needs a narrator with a point of view who offers a perspective on what happened. But you can't narrate if you don't know who you are. Many of Shear's therapy techniques are about learning to narrate in the face of great pain and devastating losses. Start with the grief diary, which records the emotional story of your everyday life. Follow that by imaginal revisiting, akin to a wide-angle shot in cinema, which helps organise a story arc amidst intense emotion.

Plotting out the story restores the narrator and the narrative. Then, you can begin to imagine a new story, a new plot for yourself. It's not a choice between grief or living, remembering or forgetting, the way Muldberg once worried it was. The book of life is a multi-volume set. A sequel can only start when the first volume is brought to a close and when the narrator knows she's going to be all right.

This story was first published on 15 November 2016
by Wellcome on mosaicscience.com

The mirror cure for phantom limb pain

■ Srinath Perur

One of the few Khmer words Stephen Sumner knows is chhue. It means 'pain', and it's something Cambodian people know a lot about from their three-decade-long civil war. Stephen, 53, is a brawny Canadian with an ebullient, even boisterous, manner. This is his third time here in as many years. He rides around on a longtail bicycle with a stack of lightweight mirrors behind the saddle, going to villages, hospitals and physical rehabilitation centres looking for people who have lost their limbs.

Just as the pain of war lingers long after it is over, so an amputee's pain can persist long after the limb has gone. It can be harrowing and difficult to treat with medication or surgery. Stephen helps people deal with their phantom pain, and he does it with mirrors.

We're in Spean Tomneap village in the Battambang province of north-western Cambodia – the most heavily mined region in one of the most heavily mined countries in the world. We've driven up along a mud road lined by fields and houses surrounded by tangled greenery. Stephen is perched on the landing near the staircase of a weathered wooden house on stilts. Chickens scurry about. A few onlookers gather.

In front of Stephen on an upturned pail sits Ven Phath, a soft-spoken, middle-aged father of five. His left trouser leg is rolled up to reveal a stump below the knee, the result of stepping on a mine in 1983. A plasticky prosthetic leg lies beside him.

Ven Phath still experiences pain in his missing foot, and Stephen is showing him how to position a mirror against the inside of his left leg, so the reflection of the right makes it look like both are still intact. 'Look. Move. Imagine,' Stephen instructs through an interpreter.

After a couple of minutes of watching his virtual left foot moving, as if revving an imaginary accelerator, Ven Phath smiles and looks up. He says he feels better already. 'Tell him,' Stephen says to the interpreter, 'if you do this twice a day, ten minutes per session, for five weeks, then chhub chhue.' Pain stop.

§ § §

'Rock and roll saved my life,' Stephen likes to say. On a balmy June evening in 2004, he was riding a scooter down a quiet country road in Tuscany when a motorist crashed into him and flung him off. Stephen has no recollection of the impact, but tyre tracks were found at the site and an anonymous phone call was placed to the Italian police about a man having gone down at Pian del Lago, on the outskirts of Siena.

The police didn't follow it up, and the area was inhabited mostly by shepherds, who tended to turn in early. When Stephen was finally found by a shepherd's nephew – returning home from band practice – he had been lying in a field for four-and-a-half hours. He only vaguely remembers feeling mud at the base of his skull, and finding it strange that he was so cold despite it being summer.

Stephen was taken to a Siena hospital with broken ribs and collarbone, a crushed arm and leg. The intensive care doctor on duty happened to be one of Stephen's English students – but she didn't recognise him until he was cleaned up. When he emerged from coma five days later, she was holding his hand. 'Don't look down,' she said. 'You've lost something.' Stephen's left leg had been amputated six inches above the knee. The doctors managed to save his arm, piecing it together with two metal plates and 28 screws.

As Stephen recuperated in hospital, he knew the leg was gone, and received regular, gruesome visual reminders when the stump was cleaned and drained.

Yet still, he felt the leg.

It began in his dreams: he recalls a particularly vivid one in which he was lying on his back on a wooden cart, his left leg visible till just above the knee. The rest of the leg was hanging down through a gap in the slats, swinging in time to the lurching of the cart. As he moved through an arid landscape, Stephen began to worry about rocks striking his foot.

Then the missing leg began to make its presence felt in Stephen's waking hours. Though he perceived it as having some movement, it was usually bent backwards at the knee. At one point, he asked the doctors if a hole could be cut in the hospital bed so his 'leg' could hang free.

During his third week in hospital, Stephen experienced an episode of pain in his phantom leg – although 'pain' turns out to be a wholly inadequate word for what he describes: excruciatingly clenched toes, jolts that he likens to being shocked by a cattle prod, writhing so violent that his head was banging against the metal sides of the hospital bed. Stephen was left in tears after the bout.

A doctor had reassured him that he wouldn't feel any phantom pain. 'You lied to me,' Stephen told him, but the doctor said they were just twinges caused by his body adjusting to change, and that they would go away. 'They didn't go away at all,' Stephen says. He returned to Canada, went through physiotherapy, began wearing a prosthetic leg and resumed his life. But the pain returned at intervals, sometimes not letting up for days at a time. 'Everything was good,' he says. 'But my leg that's not there was killing me.'

§ § §

The vivid sensation of a missing limb is experienced by almost everyone who has had a limb amputated. Lord Nelson, the naval commander who lost his right arm in battle, declared that the phantom sensations in his missing right arm were proof of the existence of a soul. Today we needn't take recourse to the mystical, since we know that the brain holds maps of the body that can be independent of the body parts it represents.

In his book *Phantoms in the Brain*, the neuroscientist Dr V S Ramachandran describes a woman born without arms who reported having phantom hands, which she used to gesticulate when she spoke. Other parts of the body have been known to come in phantom versions too: women who have undergone mastectomies report phantom breasts; people whose bladders have been removed still feel the strong urge to go; men who have undergone penectomies report phantom erections. One night, years after his leg was gone, Stephen woke up at 4am, swung his phantom leg off the bed, and crashed to the floor on his stump, necessitating a bloody trip to hospital. Even today, ten years after his amputation, he can feel a sore he had on his heel from a cycling shoe.

Phantom limbs might be a strange, even occasionally reassuring, phenomenon if they didn't hurt so much. Somewhere between 50 and 80 per cent of all amputees complain of pain in their phantom limbs. In the past, a few doctors believed phantom limbs were a form of wishful thinking on the part of amputees, and that pain in the missing limb had to be psychological in origin. But most believed the pain was caused by damaged nerves near the stump. So they tried to treat phantom pain by shortening the stump, which sometimes gave relief, but seldom for long.

Then, in the early 1990s, Dr Ramachandran and his colleagues at the University of California, San Diego, conducted simple experiments with amputees that changed the understanding of phantom limbs and sensations. When they stroked the left side of the face of a young man who had recently lost his left arm, he felt sensations not only on his face but also on his phantom hand.

They already knew that the brain's cortex has superimposed on it a virtual map of the body corresponding to sensory inputs from different parts – and that that the face's representation on this map is adjacent to the hand's. Could the young man's phantom sensations be the result of sensory inputs from his face 'invading' the now-deprived region of the cortex that mapped to his missing hand?

Brain imaging confirmed this was the case. Other researchers found that these rewired inputs might be activating neural pain pathways for the missing hand, or at least generating 'junk' signals that were interpreted by the brain as a range of sensations – including pain.

It was possible, too, that when signals sent to move the missing hand didn't lead to any corresponding visual or sensory confirmation of the movement, this dissonance was perceived as pain. The brain is known to emphasise visual feedback over other types – which may also be the reason why passengers get carsick more often than drivers. (When a passenger reads in a car moving along a curvy road, the balance sensors in the inner ear report motion that differs from what the eyes are seeing, and the dissonance is thought to be expressed as nausea. Whereas for the driver, the balance sensors in the ears, the spatial sensors in the body and what the eyes report are all in reassuring agreement.)

Dr Ramachandran and his group wondered if visual feedback of the phantom limb's movement might help relieve pain in it. They put together what they called the 'mirror box' – a simple but ingenious contraption that hid the stump while allowing a reflection of the intact limb to be superimposed over the phantom limb. Now, if the amputee moved the intact and phantom limbs in sync, the brain could 'see' the phantom limb move.

The first amputee to try the mirror box reported being able to move his phantom limb for the first time in over a decade, and he felt immediate relief from pain. Subsequent users too found they could manage their phantom pain using the box.

Surgery and medication have been found to be only slightly or not at all effective when dealing with phantom pain. Stephen knew this, and he tried to will his phantom pain away: 'Optimism. Mind over matter. I thought I could beat it.' But it kept coming back, and it kept getting worse. 'Then I tried to drink it to death, which was costly and messy in every conceivable sense, plus totally ineffective.'

In 2008, Stephen was working as a property manager in south Baja, Mexico, when he had a particularly agonising bout of phantom pain. 'I was not presentable for 72 hours,' he says. He was aware of mirror therapy from having looked online for treatments, and he decided to give it a try. He got into his truck and drove two-and-a-half hours to the nearest Home Depot to buy a mirror. He tried it right there in the parking lot, and in five minutes the pain was gone.

Stephen used the mirror for two weeks, then stopped because the pain had not returned. About a year and a half later, he felt the pain again, and this time he stayed the course for the full five weeks. He hasn't had phantom pain for over four years. 'It's gone now,' he says. 'It's gone because I treated myself with a mirror.'

§ § §

Asked to list the jobs he's held, Stephen comes up with more than 20. He's been a sailor on a commercial fishing boat, an English teacher in Italy and Saudi Arabia, a model, a screenwriter (he co-wrote the script for a German feature film about his own life, called Phantomschmerz) and an actor (involutedly playing body double for the actor portraying him in the film). He's done competitive cycling, been in construction, demonstrated cookware at a departmental store, been 'the pool guy' in Mexico and been a flight logistics officer in Afghanistan.

In the autumn of 2010, Stephen was living in a basement apartment in Vancouver when it struck him that his calling might be mirror therapy. He'd go where there were amputees in pain, give them a mirror and teach them how to use it. Cambodia was his first destination because it had an inordinately high number of amputees, and it was small and flat, which was important because Stephen was planning to bicycle with his mirrors.

And so, more than three years later, on the first day of 2014, we meet at the entrance to the rather run-down Paris Hotel in Battambang city. Stephen arrives on his bicycle, and grins as he walks towards me with hand outstretched. He's a big, strong man who might pass off as Steven Seagal if he had darker hair and was capable of a sterner mien.

His gait is somewhat lopsided – the result, I later learn, of an experimental, low-cost prosthetic knee he's testing, and the fact that this knee is tuned for cycling not walking. It is visible just below the hem of his shorts, continuing downward as a single steel pylon, without the cosmetic covering that makes prosthetics look like natural legs. This stands out even in a country with such a high proportion of amputees. This is partly by design: the success of his work depends on other 'amps' – as he affectionately calls them – accepting him as one of their own. The cycling, too, is part of winning their trust: 'It's salesmanship. It impresses people that I roll up on a bicycle.'

I witness this bonding the next morning outside the Regional Physical Rehabilitation Centre run by the International Committee of the Red Cross (ICRC). About a dozen people are gathered in the space outside a small shop – playing cards, knitting. Stephen, who seems twice the size of most people here, walks into their midst like a giant in a particularly amiable mood. 'Whose leg is that?' he asks loudly, pointing at a detached prosthetic. He sits down on a bench next to another amp, grabs the man's rudimentary, footless prosthetic and lifts it up to examine it. He smells the splintered peg at its end and recoils, only half in mock disgust. The man is overjoyed (and probably a little drunk), and he grabs Stephen's pylon.

An empty wheelchair is parked at the entrance to the centre, near scattered footwear and an ICRC-issued leg with a flip-flop on its foot. (Red Cross legs are designed with a gap between first and second toes for flip-flops.) 'Whose wheelchair is this?' Stephen asks. It belongs to a young woman in a pink skirt who's playing cards. 'Why can't you walk like everyone else?' Stephen chides her. She giggles. One of the men who speaks a little English says, 'She's just lazy.'

The Red Cross centre at Battambang fits prostheses and conducts rehabilitation free of charge to all amputees who come here. When Stephen first visited, the manager told him none of their amputees suffered from phantom pain. According to Stephen, phantom pain denial is common in Cambodia: 'Nobody wants to be thought crazy.'

Telling them his own story often helps. Stephen says he asked to speak to the amps at the ICRC centre, and 44 of them gathered in a room. Though the manager was sceptical, he agreed to translate. Through him, Stephen told the amps about his own accident. 'I told them that ten years ago I was hit, and for four years I suffered the most excruciating pain.' And then he told them how he cured himself. 'Now, how many of you have phantom limb pain?' he asked. Thirty-seven out of the 44 raised their hands. Stephen conducted a mirror therapy workshop for the centre's therapists, and left behind mirrors for them to use.

In his last two trips to Cambodia, Stephen has distributed around 600 mirrors. These are lightweight mirrors of his own design, made for him in Phnom Penh. They come in two sizes, for leg and arm amputees, and are made of an acrylic sheet taped over a silvered layer pasted on a rectangle of plastic. (Stephen originally considered tin and glass mirrors, but rejected them because of their weight and potential to injure people who could do without further injury.) Some of the mirrors went to individual amps, and some went to organisations like the ICRC or the Trauma Care Foundation.

The amps in parts of Battambang province tend to be former Khmer Rouge soldiers, though many fought on different sides of the conflict at different times. When Stephen first learned that some of the people he was treating had fought for the Khmer Rouge, he found himself in a moral quandary about coming all this way 'to help the bad guys'. But then he thought, 'You know what? Pain is pain.'

§ § §

Stephen's most satisfying day as a mirror therapist came in March 2012. He was doing a workshop with the Catholic charity Caritas at Samlout village, an 80 km ride from Battambang past palms, paddy fields, lotus ponds and houses with spirit shrines in their yards. He arrived and began treating one amp. And then: 'I look behind me and there are amputees coming from all directions.' He estimates he treated 30 people in that one memorable day, almost all of them former Khmer Rouge soldiers.

Though an atheist himself, one of Stephen's favourite collaborators is the Apostolic Prefecture of Battambang. The prefect is a Spanish Jesuit priest known, for his many years of work among amputees, as the Bishop of the Wheelchairs. It can be hard for amputees to find acceptance and employment in regular village life, so the prefecture has set up communities populated entirely by amputees and their families. A volunteer drives us one morning to Ratanak Mondol, where there's one such community.

We pass signs marking areas where mines have been cleared, signs where mines are still being cleared, and grisly hoardings warning of the dire consequences of playing with a mine. The Cambodian army, the Vietnamese army and the Khmer Rouge are estimated between them to have laid around 10 million mines in the country, which is about two mines for every three Cambodians. The mines can be Chinese, Russian, US or Vietnamese in make, reflecting the complex geopolitics that have played out here. In the more than 30 years since the war, only around half are estimated to have been recovered.

In Cambodia, landmines and unexploded ordnance killed around 20,000 people and injured 44,000 more between 1979 and 2011. Despite public information drives and demining programmes, it is still not unheard of for farmers to step on anti-personnel mines in the fields. Here in Ratanak Mondol, a tractor detonated an anti-tank mine in 2012, killing seven members of a family. Stephen says the hardest sight for him is an arm amputee: usually, it means a child has tried to play with a mine, or that the amp was trying to salvage scraps of metal to sell from it.

The prefecture's community is on a plain surrounded by low hills, with no other settlement in sight. Eight families, each with one or more amputees, live here. Each has a patch of land to farm and a bamboo-and-wood stilt house. The elevation of the stilt house is protection against floods that are common in the monsoon; for the rest of the year, the cool space under the house is its main living area.

From beneath one of the houses, four children aged four to six spill out into the yard. The eldest gets out a bicycle and scissors it perilously round the house, with a screeching younger sibling on the rear saddle. A bright heap of corncobs dries in the sun. The children's grandfather, in his 50s, is shovelling in a dry, ploughed-up field.

Only when he comes closer does it become evident that one of his legs is a prosthetic. Then a young man named Untac appears. It sounds like any other Cambodian name to us, but the volunteer from the prefecture tells us he's named after the United Nations Transitional Authority in Cambodia, the peacekeeping force that entered the country in 1992.

The community's schoolteacher is a woman of around 30. Her class has 16 children between three and six, who sing a Khmer song complete with actions to welcome us. One of the teacher's legs is a prosthetic, and she learns from Stephen how to use a mirror.

A couple of kilometres away, we visit a woman named Jian with close-cropped white hair, who has lost both her legs close to her hips. Her son is an amputee too. She rolls up on her wheelchair and slides off onto the platform of her bamboo hut to welcome us. She's fine except for an eye that hurts, she tells the volunteer, flashing a betel-red smile.

Stephen has tried to treat phantom pain in bilateral amputees by having small-built people sit in the amp's lap and move their legs. While this might sound like a long shot, the brain does appear easily fooled into adopting limbs as its own. Once, Stephen 'sprains' his prosthetic's ankle and, to fix it, whips up the leg the wrong way so the sole of his left shoe is in front of his face. I almost howl in pain even though it's only a prosthetic leg – and it's not even mine.

Over breakfast at a café in Battambang city, I ask Stephen what phantom pain feels like. 'It's electric,' he says, 'like this' –

and jerks so violently that the heavy table we're sitting at rises a couple of inches and a glass of water topples over. 'There's also burning and crushing, but the worst is the itching. I could kill someone.'

The most important thing to understand about phantom limb pain, he says, is that 'It's not in the head, it's in the limb.' But he also feels there's a psychosomatic component to its being relieved by mirror therapy. 'When I finally tried mirror therapy on myself,' he explains, 'it almost had to work. I mean, I needed something.'

§ § §

There isn't consensus among neuroscientists, doctors and therapists about the mechanism behind phantom limb pain, and how – or even if – mirror therapy addresses it. For the last 20 years or so, it has been widely thought that phantom limb pain is mainly a consequence of 'central maladaptive reorganisation': changes in and around the missing limb's corresponding patch of the cortex. According to this theory, phantom pain follows intruding sensory inputs from areas of the cortex that are next to the deprived area representing the missing limb.

In 2013, however, neuroscientist Dr Tamar Makin of the University of Oxford and her colleagues published results that questioned this understanding. The maladaptive reorganisation idea had previously been supported by looking for the intrusion it implied, for instance from the area of the cortex representing the face to the adjacent area representing the hand. Dr Makin took a different approach, looking at the hand area in the cortex that was supposedly being encroached upon.

She found that when amputees moved their phantom hands, as some can, the representation of the hand in the cortex lit up on a brain scan, seemingly well-preserved. This was contrary to the accepted notion that the missing hand's piece of cortex wasn't getting any sensory inputs. Further, the more the reported phantom pain, the better the missing hand's representation was preserved. This brings up the possibility that phantom limb pain may somehow be involved in maintaining the brain's representation of missing limbs.

Though Dr Makin's work did not directly address the question, she speculated in an email interview that the main cause of phantom limb pain is 'junk' inputs from peripheral nerves near the stump itself. If that were the case – the standard objection to this theory goes – then treatment near the stump should cure phantom limb pain, but it usually doesn't. Dr Makin argues that the effect of an amputation is felt not only at the endings of nerves but also in their cell bodies, close to the spine. So it's not possible to simply 'shut down' those peripheral nerves by localised treatment near the stump.

Dr Makin is sceptical of mirror therapy – 'I don't believe in magic,' she says – and feels that the relief felt by many amputees is probably explained by the placebo effect. If mirror therapy does any more than that, she suggests, it would be by increasing normal inputs to the parts of the brain associated with the missing limb, thus 'diluting' junk inputs.

In contrast, Dr Herta Flor, a neuroscientist at the University of Heidelberg, says of mirror therapy: 'It definitely works, though not in all patients. It works by providing visual feedback to the brain about a functional arm rather than a missing limb, and this changes the central maladaptive reorganisation back to normal.' Dr Flor and her colleagues were the first, in 1995, to report neuroimaging studies that linked phantom pain to central sensory reorganisation in the cortex. According to her, the central versus peripheral debate is something of a battle of straw men because the two are not mutually exclusive: computer modelling predicts that aberrant input from the periphery can lead to central changes. She says, 'I personally believe that both – peripheral input and central changes – contribute to phantom limb pain.'

Several controlled trials of mirror therapy have shown it works better than placebo for treating phantom pain, paralysis and complex regional pain syndrome. But a 2011 meta-analysis found some of the studies to be of poor quality and could not reach a definite conclusion about the efficacy of mirror therapy.

'Patients are complicated and nothing works for everybody,' says Dr Eric Altschuler of the New Jersey Medical School, Rutgers University, and a collaborator of Dr Ramachandran. In his experience, there is more than one kind of phantom pain, a distinction that's not often taken into account in trials. He says, 'The mirror works best for immobile or clenched phantoms. It doesn't necessarily work for burning pain.' Even so, he adds, 'the mirror is the only effective treatment'. As for the cause of phantom pain, he feels that reorganisation in the brain is behind it.

When he teaches mirror therapy, Stephen offers a simplified explanation of the brain reorganisation theory. Pointing to his head, he says, 'You have a commander here that controls the body.' Many of the people he treats have been soldiers, and they are familiar with talk of commanders and maps.' The commander has a map of the whole body. When the map doesn't match the body, the commander panics and you feel pain. This mirror is to trick the commander into thinking the leg still exists, so he stops panicking and the pain goes away.'

From personal experience and from working with hundreds of amputees, Stephen has no doubt that the therapy helps those with phantom pain. In that light, he isn't overly affected by neuroscience debates. 'I don't have too much time for party-crashers,' he says. 'I got amps to see.'

§ § §

Whatever the science of it, there's something marvellously loopy about a one-legged man on a bicycle riding into villages with a bunch of mirrors. I set out with Stephen early one morning after a cup of thick, strong coffee sweetened with condensed milk. We're riding to Samlout.

An hour after we've started, Stephen drifts across the road to stop near a dilapidated temple. Bun Thoeun, 59, works under a tree here as a bicycle mechanic. He fought for the Khmer Rouge from 1979 to 1983, when he lost his leg. Then he spent ten years in an international relief camp on the Thai border, where he learned some English. He's delighted to see Stephen, and swings over surprisingly quickly on his one good leg and a crutch. Stephen's bike doesn't really need air, but Bun Thoeun insists it does.

A little beyond Samlout is the rehabilitation workshop run by the Trauma Care Foundation. It's run almost entirely by amputees, who make wheelchairs, crutches, walkers and prosthetics for other amps. If it weren't for this workshop, those who live here would have to go to the ICRC centre at Battambang to get prostheses fitted or modified. Given the cost of going there, and the lost earnings, many choose to get by on crutches or crude prostheses of their own fashioning. The workshop's funding is due to run out later this year, as the foundation moves on to other conflict zones.

Seven or eight people whom Stephen treated on his previous visit have gathered. One says he had pain in his phantom foot but feels much better now after using the mirror. Another used to be able to tell the coming of rain from a clenching near his stump, but not any more. 'For how many of you did the pain disappear?' Stephen asks. Almost all present raise their hands, and there's clapping in the workshop.

Stephen had trained one of the men here to be a mirror therapist, and left him a stack of mirrors. Now he holds a refresher class, using a whiteboard to write down how often to use the mirror. It's best, he emphasises, to recruit an amputee's spouse or child to remind the amp to use the mirror every day for five weeks. 'I could stay here in Battambang all my life and I'd never run out of amps,' Stephen says. But he will soon move on to other amps in other places – Laos and northern Sri Lanka on this trip.

Even where Stephen has trained people and left mirrors, clinics have sometimes been slow to use them. At the ICRC centre in Battambang, he had to visit again, locate his mirrors (atop a storage locker), wipe off the dust and exhort people to use them. But his persistence is paying dividends – last year he gave a workshop in Battambang for an international panel of Red Cross therapists working in various parts of the world. He's set to do another soon.

One reason for the slow embrace of mirror therapy in the field may be the whiff of scepticism that still surrounds it. As well as the doubts from some neuroscientists, Stephen routinely has doctors and therapists telling him, 'Well, it's not scientific' – simply because mirror therapy looks too simple.

'I am an above-knee amp. I cured myself with a mirror,' Stephen counters. 'I challenge someone in a white lab coat who has never been anywhere where it hurts to tell me otherwise.'

He also thinks there is a systemic lack of enthusiasm about mirror therapy because there's no money to be made from it. 'It would be much better,' he says wryly, 'to have a clinical frequent flyer with an escalating OxyContin habit. Works for all parties except the poor amp.'

Stephen has been resistant to steady employment all his life. To pay his way as a mirror therapist, he has worked temporary jobs in Canada, saving up for the mirrors and the travel expenses. But he's exhausted now, and doesn't know how much longer he can keep this up.

One evening, after a couple of beers, the tiredness and frustration bubble briefly to the surface. 'I can't believe no one else is doing this,' Stephen says to me, pounding the table with his fist. 'It's super-effective. I'd have thought there would be thousands of people riding around with mirrors, but there are not... What is wrong with people?'

This story was first published on 8 July 2014
by Wellcome on mosaicscience.com

Can you think yourself into a different person?

■ Will Storr

For years she had tried to be the perfect wife and mother but now, divorced, with two sons, having gone through another break-up and in despair about her future, she felt as if she'd failed at it all, and she was tired of it. On 6 June 2007 Debbie Hampton, of Greensboro, North Carolina, took an overdose. That afternoon, she'd written a note on her computer: 'I've screwed up this life so bad that there is no place here for me and nothing I can contribute.' Then, in tears, she went upstairs, sat on her bed, and put on a Dido CD to listen to as she died.

But then she woke up again. She'd been found, rushed to hospital, and saved. 'I was mad,' she says. 'I'd messed it up. And, on top of that, I'd brain-damaged myself.' After Debbie emerged from her one-week coma, her doctors gave her their diagnosis: encephalopathy. 'That's just a general term which means the brain's not operating right,' she says. She couldn't swallow or control her bladder, and her hands constantly shook. Much of the time, she couldn't understand what she was seeing. She could barely even speak. 'All I could do was make sounds,' she says. 'It was like my mouth was full of marbles. It was shocking, because what I heard from my mouth didn't match what I heard in my head.' After a stay in a rehabilitation centre, she began recovering slowly. But, a year in, she plateaued. 'My speech was very slow and slurred. My memory and thinking was unreliable. I didn't have the energy to live a normal life. A good day for me was emptying the dishwasher.'

It was around this time that she tried a new treatment called neurofeedback. She was required to have her brain monitored while playing a simple Pac-Man-like game, controlling movements by manipulating her brain waves. 'Within ten sessions, my speech improved.' But Debbie's real turnaround happened when her neurofeedback counsellor recommended a book: the international bestseller The Brain that Changes Itself by Canadian psychotherapist Norman Doidge. 'Oh my God,' she says. 'For the first time it really showed me it was possible to heal my brain. Not only that it was possible, that it was up to me.'

After reading Doidge's book, Debbie began living what she calls a 'brain-healthy' life. That includes yoga, meditation, visualisation, diet and the maintenance of a positive mental attitude. Today, she co-owns a yoga studio, has written an autobiography and a guide to 'brain-healthy living' and runs the website thebestbrainpossible.com. The science of neuroplasticity, she says, has taught her that, 'You're not stuck with the brain you're born with. You may be given certain genes but what you do in your life changes your brain. And that's the magic wand.' Neuroplasticity, she says, 'allows you to change your life and make happiness a reality. You can go from being a victim to a victor. It's like a superpower. It's like having X-ray vision.'

Debbie's not alone in her enthusiasm for neuroplasticity, which is what we call the brain's ability to change itself in response to things that happen in our environment. Claims for its benefits are widespread and startling. Half an hour on Google informs the curious browser that neuroplasticity is a 'magical' scientific discovery that shows that our brains are not hardwired like computers, as was once thought, but like 'play-doh' or a 'gooey butter cake'. This means that 'our thoughts can change the structure and function of our brains' and that by doing certain exercises we can actually, physically increase our brain's 'strength, size and density'. Neuroplasticity is a 'series of miracles happening in your own cranium' that means we can be better salespeople and better athletes, and learn to love the taste of broccoli. It can treat eating disorders, prevent cancer, lower our risk of dementia by 60 per cent and help us discover our 'true essence of joy and peace'. We can teach ourselves the 'skill' of happiness and train our brains to be 'awesome'. And age is no limitation: neuroplasticity shows that 'our minds are designed to improve as we get older'. It doesn't even have to be difficult. 'Simply by changing your route to work, shopping at a different grocery store, or using your non-dominant hand to comb your hair will increase your brain power.' As the celebrity alternative-medicine guru Deepak Chopra has said, 'Most people think that their brain is in charge of them. We say we are in charge of our brain.'

Debbie's story is a mystery. The techniques promising to change her brain via an understanding of the principles of neuroplasticity have clearly had tremendous positive effects for her. But is it true that neuroplasticity is a superpower, like X-ray vision? Can we really increase the weight of our brain just by thinking? Can we lower our risk of dementia by 60 per cent? And learn to love broccoli?

Some of these seem like silly questions, but some of them don't. That's the problem. It's hard, for the non-scientist, to understand what exactly neuroplasticity is and what its potential truly is. 'I've seen tremendous exaggeration,' says Greg Downey, an anthropologist at Macquarie University and co-author of the popular blog Neuroanthropology. 'People are so excited about neuroplasticity they talk themselves into believing anything.'

§ § §

For many years, the consensus was that the human brain couldn't generate new cells once it reached adulthood. Once you were grown, you entered a state of neural decline. This was a view perhaps most famously expressed by the so-called founder of modern neuroscience, Santiago Ramón y Cajal. After an early interest in plasticity, he became sceptical, writing in 1928, 'In adult centres the nerve paths are something fixed, ended, immutable. Everything may die, nothing may be regenerated. It is for the science of the future to change, if possible, this harsh decree.' Cajal's gloomy prognosis was to rumble through the 20th century.

Although the notion that the adult brain could undergo significant positive changes received sporadic attention, throughout the 20th century, it was generally overlooked, as a young psychologist called Ian Robertson was to discover in 1980. He'd just begun working with people who had had strokes at the Astley Ainslie Hospital in Edinburgh, and found himself puzzled by what he was seeing. 'I'd moved into what was a new field for me, neuro-rehabilitation,' he says. At the hospital, he witnessed adults receiving occupational therapy and physiotherapy. Which made him think... if they'd had a stroke, that meant a part of their brain had been destroyed. And if a part of their brain had been destroyed, everyone knew it was gone for ever. So how come these repetitive physical therapies so often helped? It didn't make sense. 'I was trying to get my head around, what was the model?' he says. 'What was the theoretical basis for all this activity here?' The people who answered him were, by today's standards, pessimistic.

'Their whole philosophy was compensatory,' Robertson says. 'They thought the external therapies were just preventing further negative things happening.' At one point, still baffled, he asked for a textbook that explained how it all was supposed to work. 'There was a chapter on wheelchairs and a chapter on walking sticks,' he says. 'But there was nothing, absolutely nothing, on this notion that the therapy might actually be influencing the physical reconnection of the brain. That attitude really went back to Cajal. He really influenced the whole mindset which said that the adult brain is hardwired, all you can do is lose neurons, and that if you have brain damage all you can do is help the surviving parts of the brain work around it.'

But Cajal's prognosis also contained a challenge. And it wasn't until the 1960s that the 'science of the future' first began to rise to it. Two stubborn pioneers, whose tales are recounted so effectively in Doidge's bestseller, were Paul Bach-y-Rita and Michael Merzenich. Bach-y-Rita is perhaps best known for his work helping blind people 'see' in a new and radically different way. Rather than receiving information about the world from the eyes, he wondered if they could take it in in the form of vibrations on their skin. They'd sit on a chair and lean back on a metal sheet. Pressing up against the back side of that metal sheet were 400 plates that would vibrate in accord with the way an object was moving. As Bach-y-Rita's devices became more sophisticated (the most recent version sits on the tongue), congenitally blind people began to report having the experience of 'seeing' in three dimensions. It wasn't until the advent of brain-scanning technology that scientists began to see evidence for this incredible hypothesis: that information seemed to be being processed in the visual cortex. Although this hypothesis is yet to be firmly established, it seems as if their brains had rewired themselves in a radical and useful way that had long been thought impossible.

Merzenich, meanwhile, helped to confirm in the late 1960s that the brain contains 'maps' of the body and the outside world, and that these maps have the ability to change. Next, he co-developed the cochlear implant, which helped deaf people hear. This relies on the principle of plasticity, as the brain needs to adapt to receive auditory information from the artificial implant instead of the cochlea (which, in the deaf person, isn't working). In 1996 he helped establish a commercial company that produces educational software products called Fast ForWord for 'enhancing the cognitive skills of children using repetitive exercises that rely on plasticity to improve brain function,' according to their website. As Doidge writes, 'In some cases, people who have had a lifetime of cognitive difficulties get better after only thirty to sixty hours of treatment.'

Although it took several decades, Merzenich and Bach-y-Rita were to help prove that Cajal and the scientific consensus were wrong. The adult brain was plastic. It could rewire itself, sometimes radically. This came as a surprise to experts like Robertson, now a Director of Trinity College Dublin's Institute of Neuroscience. 'I can look back on giving lectures at Edinburgh University to students where I gave wrong information, based on the dogma which said that, once dead, a brain cell cannot regenerate and plasticity happens in early childhood but not later,' he says.

It wasn't until the publication of a series of vivid studies involving brain scans that this new truth began to be encoded into the synapses of the masses. In 1995, neuropsychologist Thomas Elbert published his work on string players that showed the 'maps' in their brain that represented each finger of the left hand – which they used for fingering – were enlarged compared to those of non-musicians (and compared to their own right hands, not involved in fingering). This demonstrated their brains had rewired themselves as a result of their many, many, many hours of practice. Three years later, a Swedish–American team, led by Peter Eriksson of Sahlgrenska University Hospital, published a study in Nature that showed, for the very first time, that neurogenesis – the creation of new brain cells – was possible in adults. In 2006, a team led by Eleanor Maguire at the Institute of Neurology at University College London found that the city's taxi drivers have more grey matter in one hippocampal area than bus drivers, due to their incredible spatial knowledge of London's maze of streets. In 2007, Doidge's The Brain that Changes Itself was published. In its review of the book, the *New York Times* proclaimed that 'the power of positive thinking has finally gained scientific credibility'. It went on to sell over one million copies in over 100 countries. Suddenly, neuroplasticity was everywhere.

§ § §

It's easy, and perhaps even fun, to be cynical about all this. But neuroplasticity really is a remarkable thing. 'What we do know is that almost everything we do, all our behaviour, thoughts and emotions, physically change our brains in a way that is underpinned by changes in brain chemistry or function,' says Robertson. 'Neuroplasticity is a constant feature of the very essence of human behaviour.' This understanding of the brain's power, he says, opens up new techniques for treating a potentially spectacular array of illnesses. 'There's virtually no disease or injury, I believe, where the potential doesn't exist for very intelligent application of stimulation to the brain via behaviour, possibly combined with other stimulation.'

Does he agree that the power of positive thinking has now gained scientific credibility? 'My short answer is yes,' he says. 'I do think human beings have much more control over their brain function than has been appreciated.' The long answer is: yes, but with caveats. First there's the influence of our genes. Surely, I ask Robertson, they still hold a powerful influence over everything from our health to our character? 'My own crude rule of thumb is a 50–50 split in terms of the influence of nature and that of nurture,' he says. 'But we should be very positive about that 50 per cent that's environmental.'

Adding extra tangle to the already confused public discussion of neuroplasticity is the fact that the word itself can mean several things. Broadly, says Sarah-Jayne Blakemore, Deputy Director of London's Institute of Cognitive Neuroscience, it refers to 'the ability of the brain to adapt to changing environmental stimuli'. But the brain can adapt in many different ways. Neuroplasticity can refer to structural changes, such as when neurons are created or die off or when synaptic connections are created, strengthened or pruned. It can also refer to functional reorganisations, such as those experienced by the blind patients of Paul Bach-y-Rita, whose contraptions triggered their brains to start using their visual cortices, which had previously been redundant.

On the larger, developmental scale, there are two categories of neuroplasticity. They are 'really different,' says Blakemore. 'You need to differentiate between them.' Throughout childhood our brains undergo a phase of 'experience-expectant' plasticity. They 'expect' to learn certain important things from the environment, at certain stages, such as how to speak. Our brains don't finish developing in this way until around our mid-20s. 'That's why car insurance premiums are so high for people under 25,' says Robertson. 'Their frontal lobes aren't fully wired up to the rest of their brains until then. Their whole capacity for anticipating risk and impulsivity isn't there.' Then there's 'experience-dependent' plasticity. 'That's what the brain does whenever we learn something, or whenever something changes in the environment,' says Blakemore.

One way in which science has been exaggerated has been by the blending of these different types of change. Some writers have made it seem as if almost anything counts as 'neuroplasticity', and therefore revolutionary and magical and newsworthy. But it's definitely not news, for example, that the brain is highly affected by its environment when we're young. Nevertheless, in The Brain that Changes Itself Norman Doidge observes the wide variety of human sexual interests and calls it 'sexual plasticity'. Neuroscientist Sophie Scott, Deputy Director of London's Institute of Cognitive Neuroscience, is dubious. 'That's just the effect of growing up on your brain,' she says. Doidge even uses neuroplasticity to explain cultural changes, such as the broad acceptance in the modern age that we marry for romantic love, rather than socioeconomic convenience. 'That isn't neuroplasticity,' says Scott.

This, then, is the truth about neuroplasticity: it does exist, and it does work, but it's not a miracle discovery that means that, with a little effort, you can turn yourself into a broccoli-loving, marathon-running, disease-immune, super-awesome genius. The 'deep question', says Chris McManus, Professor of Psychology and Medical Education at University College London, is, 'Why do people, even scientists, want to believe all this?' Curious about the underlying causes of the neuroplasticity craze, he believes it is just the latest version of the personal-transformation myth that's been haunting the culture of the West for generations.

§ § §

'People have all sorts of dreams and fantasies and I don't think we're very good at achieving them,' says McManus. 'But we like to think that when somebody is unsuccessful in life they can transform themselves and become successful. It's Samuel Smiles, isn't it? That book he wrote, Self-Help, was the positive thinking of Victorian times.'

Samuel Smiles [Full disclosure: Samuel Smiles is my great-great-uncle] is commonly cited as the inventor of the 'self-help' movement and his book, just like Doidge's, spoke to something deep in the population and became a surprise bestseller. The optimistic message Smiles delivered spoke of both the new, modern world and the dreams of the men and women living in it. 'In the 18th century, power had all been about the landed gentry,' says historian Kate Williams. 'Smiles was writing in the era of the Industrial Revolution, widespread education and economic opportunities offered by Empire. It was the first time a middle-class man could work hard and do well. They needed a formidable work ethic to succeed, and that's what Smiles codified in Self-Help.'

In the latter part of the 19th century, US thinkers adapted this idea to reflect their national belief that they were creating a new world. Adherents of the New Thought, Christian Science and Metaphysical Healing movements stripped away much of the talk of hard work, insisted upon by the Brits, to create the positive thinking movement to which some believe neuroplasticity has given scientific credence. Psychologist William James called it 'the mind-cure movement', the 'intuitive belief in the all-saving power of healthy-minded attitudes as such, in the conquering efficacy of courage, hope, and trust, and a correlative contempt for doubt, fear, worry, and all nervously precautionary states of mind'. Here was the inherently American notion that self-confidence and optimism – thoughts themselves – could offer personal salvation.

This myth – that we can be whoever we want to be, and achieve our dreams, as long as we have sufficient self-belief – emerges again and again, in our novels, films and news, and TV singing competitions featuring Simon Cowell, as well as unexpected crazes like that for neuroplasticity. One previous, and remarkably similar, incarnation was Neuro-Linguistic Programming, which had it that psychological conditions such as depression were nothing more than patterns learned by the brain and that success and happiness were just a matter of reprogramming it. The idea appeared in a more academic costume, according to McManus, in the form of what's known as the Standard Social Science Model. 'This is the idea from the 1990s where, in effect, all human behaviour is infinitely malleable and genes play no role at all.'

But the plasticity boosters have an answer to the tricky question of genes, and their heavy influence over all matters of health, life and wellbeing. Their answer is epigenetics. This is the relatively new understanding of the ways in which the environment can change how genes express themselves. Deepak Chopra has said that epigenetics has shown us that, 'regardless of the nature of the genes we inherit from our parents, dynamic change at this level allows us almost unlimited influence on our fate'.

Jonathan Mill, Professor of Epigenetics at the University of Exeter, dismisses this kind of claim as 'babble'. 'It's a really exciting science,' he says, 'but to say these things are going to totally rewire your whole brain and gene functioning is taking it far too far.' And it's not just Chopra, he adds. Broadsheet newspapers and academic journals have also been guilty, at times, of falling for the myth. 'There have been all sorts of amazingly overhyped headlines. People who have been doing epigenetics for a while are almost in despair, at the moment, partly because it's being used as an explanation for all sorts of things without any real direct evidence.'

§ § §

Just as epigenetics doesn't fulfil our culture's promise of personal transformation, nor does neuroplasticity. Even some of the more credible-sounding claims are, according to Ian Robertson, currently unjustifiable. Take the one about reducing our risk of dementia by 60 per cent. 'There is not a single scientific study that has ever shown that any intervention of any kind can reduce the risk of dementia by 60 per cent, or indeed by any percentage,' he says. 'No one has done the research using appropriate control-group methodologies to show that there is any cause-and-effect link.'

Indeed, the clinical record for many famous treatments that use the principles of neuroplasticity is notably mixed. In June 2015, the Food and Drug Administration in the US permitted the marketing of the latest iteration of Bach-y-Rita's on-the-tongue 'seeing' devices for the blind, citing successful studies. And yet a 2015 Cochrane Review of constraint induced movement therapy – a touchstone treatment for neuroplasticity evangelists that offers improvements in motor function for people who have had a stroke – found that 'these benefits did not convincingly reduce disability'. A 2011 meta-analysis of neuroplasticity God-father Michael Merzenich's Fast ForWord learning techniques, described to such thrilling effect by Doidge, found 'no evidence' that they were 'effective as a treatment for children's oral lan-guage or reading difficulties'. This, according to Sophie Scott, goes for other treatments too. 'There's been a lot of excitement about brain-training packages and, actually, big studies of those tend not to show very much effect,' she says. 'Or they show you've got better at the thing you've practised at, but it doesn't generalise to something else.' In November 2015, a team lead by Clive Ballard at King's College London found some evidence that online brain-training games might help reasoning, attention and memory in the over-50s.

It's perhaps understandable why crazy levels of hope are raised when people read tales of apparently miraculous recovery from brain injury that feature people seeing again, hearing again, walking again and so on. These dramatic accounts can make it sound as if anything is possible. But what's usually being described, in these instances, is a very specific form of neuro-plasticity – functional reorganisation – which can happen only in certain circumstances. 'The limits are partly architectural,' says Greg Downey. 'Certain parts of the brain are better at doing certain kinds of thing, and part of that comes simply from where they are.'

Another limitation, for the person hoping to develop a superpower, is the simple fact that every part of a normal brain is already occupied. 'The reason you get reorganisation after an amputation, for example, is that you've just put into unemployment a section of the somatosensory cortex,' he says. A healthy brain just doesn't have this available real estate. 'Because it keeps getting used for what it's being used for, you can't train it to do something else. It's already doing something.'

Age, too, presents a problem. 'Over time, plastic sets,' says Downey. 'You start off with more of it and space for movement slowly decreases. That's why a brain injury at 25 is a total different ballgame to a brain injury at seven. Plasticity says you start off with a lot of potential but you're laying down a future that's going to become increasingly determined by what you've done before.'

Robertson speaks of treating a famous writer and historian who'd had a stroke. 'He completely lost the capacity for all expressive language,' he says. 'He couldn't say a word, he couldn't write. He had a huge amount of therapy and no amount of stimulation could really recover that because the brain had become hyper-specialised and a whole network had developed for the highly refined production of language.' Despite what the currents of our culture might insistently beckon us towards believing, the brain is not Play-Doh. 'You can't open up new areas of it,' says McManus. 'You can't extend it into different parts. The brain isn't a mass of grey gloop. You can't do anything you like.'

Even the people whose lives are being transformed by neuroplasticity are finding that brain change is anything but easy. Take recovery from a stroke. 'If you're going to recover the use of an arm, you may need to move that arm tens of thousands of times before it begins to learn new neural pathways to do that,' says Downey. 'And, after that, there's no guarantee it's going to work.' Scott says something similar about speech and language therapy. 'There were dark days, say, 50 years ago, where if you'd had a stroke you didn't get that kind of treatment other than to stop you choking because they'd decided it doesn't work. But now it's becoming absolutely clear that it does, and that it's a phenomenally good thing. But none of it comes for free.'

Those who over-evangelise emerging disciplines like neuro-plasticity or epigenetics can sometimes be guilty of talking as if the influence of our genes no longer matters. Their enthusi-asm can make it seem, to the non-specialist, as if nurture can easily conquer nature. This is a story that attracts people in great numbers, to newspapers, blogs and gurus, because it's one our culture reinforces, and one we want to believe: that radical personal transformation is possible, that we have the potential to be whoever and whatever we want to be, that we can find happiness, success, salvation – all we need to do is try. We are dreamers down to our very synapses, we are the people of the American Dream.

Of course, it's our malleable brains that have moulded them-selves to these rhythms. As we grow up, the optimistic myths of our culture become so embedded in our sense of self that we can lose touch with the fact that they are just myths. The irony is that when scientists carefully describe the blind seeing and the deaf hearing, and we hear it as talk of wild miracles, it's the fault of our neuroplasticity.

This story was first published on 17 November 2015
by Wellcome on mosaicscience.com

How to survive a troubled childhood

■ Lucy Maddox

The landscape of the Hawaiian islands is as idyllic as a postcard: long, sandy beaches, hibiscus flowers, clear waters of tropical fish and coral reefs. When you arrive at the airport the air is warm and ukulele music is piped out at you. Flower garlands are for sale.

There are hundreds of islands in the Hawaiian archipelago, spread over 1,500 miles in the central Pacific Ocean. The eight main islands include Kauai, Maui and the island of Hawaii, nicknamed The Big Island to differentiate it from the whole state. The Big Island has a live but well-tempered volcano, which has created a dream-like landscape of black rock. Hawaiian myths explain the weird natural features including the tiny, tear-shaped lava rocks that lie all around on the volcano's sides, named 'Pele's tears' after the Hawaiian fire goddess. The legend has it that if you take any of Pele's tears away with you, you will be cursed for the rest of your life, unless you return them to where they belong. In the midst of all the beauty, Hawaii has some dark and sinister stories.

Mirena*, who is now 60, was born on the island of Kauai. I meet her on Skype: me in my sitting room in the evening, the English weather dark outside; her in the office where she works at a local school, early in the morning, the light bright and palm trees visible from the window. Mirena is a charismatic woman who speaks with passion. She comes across as warm, caring and professional, and her silver earrings flash against her dark, short hair. Mirena remembers a Hawaii from before the tourism boom, growing up playing in the red Anahola dirt, running through the cane fields. She recalls the simplicity of much of the lifestyle then, the excitement when the first stop light was erected for the cane field trucks, with children walking across the island to go and look at it.

Despite the setting, Mirena's childhood was far from a paradise. 'I saw things...' she says. 'I saw things children shouldn't see.'

§ § §

Mirena was born in 1955, the year that an experiment began. Mirena's family, like all families on Kauai who had babies in that year, was approached by two researchers: Emmy Werner and Ruth Smith. Werner and Smith were psychologists who had become interested in which factors in a child's early life set them off on a positive trajectory, and which ones really get in the way of them reaching their full potential. Little did the families or the researchers know that this would turn into one of the longest studies of child development and childhood adversity that there has ever been.

'We were not even born when the initial investigations started,' says Mirena. 'There were 698 families that said, 'Yes, we'll support whatever you need.'' The researchers monitored the families from before the babies' birth, following them and checking in at ages one, two, 10, 18, 32 and 40. They managed to track most of the cohort. 'When you come from an island such as Kauai, people don't move away,' explains Mirena. 'And if they do move away, chances are you're going to find somebody, some relative, who knows where they are... they were pretty successful in tracking us down.'

The researchers followed first the parents and then the children, finding out all sorts of things about how the cohort were doing and what sort of background they had come from. They used a mix of semi-structured interviews, questionnaires and community records of mental health, marriage, divorce, criminal convictions, school achievement and employment.

'My recollection of being a participant, I think the first time, age 18, I was already a young mother,' says Mirena. 'I got a phone call from Dr Ruth Smith... she introduced herself and said, 'Can I come and talk story?' – which is interview. We're talking story right now.'

Mirena spent her childhood in a three-bedroom house, with her parents and six siblings. The children walked the mile to and from school, arriving back home to a house they were responsible for keeping clean and tidy. She recalls the black-and-white TV with a piece of shaded paper stuck on the front to make it look like colour.

Hawaii back then was a mix of plantations and a growing hotel industry. Mirena's father worked for the coastguard. Her mother worked for Aloha Airlines as an entertainer, hula dancing and singing. Mirena's family had very little money to feed the seven children, and her father drank heavily. Her parents' marriage was often difficult and sometimes physically violent. 'We were very poor, my father was an alcoholic,' Mirena says.

The researchers in the Kauai study separated the nearly 700 children involved into two groups. Approximately two-thirds were thought to be at low risk of developing any difficulties, but about one-third were classed as 'high-risk': born into poverty, perinatal stress, family discord (including domestic violence), parental alcoholism or illness.

'Well, my family definitely fell in the 'at-risk' category,' says Mirena. 'And you know, I didn't fully... when you live in an environment, that's just where you are. You don't ever stand back and say, 'Well, I was at risk.''

The researchers expected to find that the 'high-risk' children would do less well than the others as they grew up. In line with those expectations, they found that two-thirds of this group went on to develop significant problems. But totally unexpectedly, approximately one-third of the 'high-risk' children didn't. They developed into competent, confident and caring individuals, without significant problems in adult life. The study of what made these children resilient has become as least as important as the study of the negative effects of a difficult childhood. Why did some of these children do so well despite their adverse circumstances?

§ § §

The study of how some of these Kauai children thrived despite early adversity is still ongoing. Lali McCubbin is the current principal investigator. The daughter of Hamilton McCubbin, who worked with the original researchers, she knows the history of the project well and has some Hawaiian heritage herself.

'This was a really groundbreaking study,' she says. 'What made the study unique was that despite these risk factors... that wasn't a guarantee... that you would be on a certain trajectory. And in fact, what we found was there was resilience. These children were able to thrive, were able to grow, were able to develop... able to live productive and fulfilling lives.

'A lot of these risk factors are what my father grew up with,' McCubbin adds. 'Alcoholism, strict discipline, domestic violence. And I was very fortunate, I didn't grow up with that, I had a stable home, a very loving home. None of those risk factors. So I was fascinated with how you can take a risk factor intergenerationally and create not intergenerational trauma but intergenerational resilience.'

Three clusters of protective factors tended to mark out the children who did well despite being 'high-risk': aspects of the child's temperament, having someone who was consistently caring (typically but not necessarily a family member), and having a sense of belonging to a wider group.

Overall, the third of 'high-risk' children who showed resilience tended to have grown up in families of four children or fewer, with two years or more between them and their siblings, few prolonged separations from their primary caregiver, and a close bond with at least one caregiver. They tended to be described positively as infants, with adjectives such as 'active', 'cuddly' or 'alert', and they had friends at school and emotional support outside of their families. Those who did better also tended to have more extracurricular activities and, if female, to avoid pregnancy until after their teenage years.

The picture was complex, though, with different factors seeming to be important at different ages, McCubbin explains. At age 10, doing well was linked to having been born without complications and having parents with fewer difficulties such as mental health problems, chronic poverty or trouble parenting. At age 10 and 18, positive individual personality traits seemed to help, as well as the presence of positive relationships, though not necessarily with the parents. At age 32 and 40, having a stable marriage was protective, as was participation in the armed forces.

Strikingly, even some children who had 'gone off the rails' in their teenage years managed to turn things around and get their lives back on track by the time they were in their 30s and 40s, often without the help of mental health professionals. Many of the factors involved in such turnarounds, and several of the factors associated with resilience throughout the children's lives, involve relationships of some kind, whether within the context of a larger community – a school, a religion, the armed services – or in the context of one important person.

'Our relationships really are key,' says McCubbin. 'One person can make a big difference.'

Wider research suggests that the more risk factors children face, the more protective factors they are likely to need to compensate. But as McCubbin says, 'A lot of the research supports this idea of relationships, and the need to have a sense of someone that believes in you or someone that supports you – even in a chaotic environment, just having that one person.'

'Children don't know what goes on in the lives of the adults who care for them,' says Mirena. 'They're subject to that life and not by choice. No child chooses to be poor, no child chooses to have alcoholism in their home. It just is, and you deal with it.'

Mirena has done a lot of thinking about her parents' role in her life, and the importance of having caring and supportive people and environments outside the immediate family home. 'My parents, bless their hearts, love them to pieces, but they didn't do what parents ought to do,' says Mirena. 'They were too busy trying to figure out themselves... trying to figure out what do you do with this house full of kids and not enough money to support them... My mom was too busy coping with an alcoholic husband ...'

As the eldest child, Mirena often felt responsible for trying to resolve family rows. She has memories of her parents' violent arguments. 'I saw my mom just raging with my dad. He's in the kitchen, sitting, she's busted all the bottles all over the kitchen... There's blood everywhere and I'm thinking, 'What can I do? I'm just a kid."

Mirena thinks her grandmother played a pivotal role. 'Luckily for me, we had a gran-ma down the street,' she says. 'My mother's parents lived nearby. They made a huge difference for me, just knowing that somebody loved me no matter what. And I was not always the easiest child. I was sometimes very aggressive and you become that when you have to defend your family. And we spent most of our days outside, so dirty, we were always dirty. Long, tangled hair.

'When things were really bad I would end up at my gran-ma's house. She was not living that far away... I cut through the park and cut through the cane fields and by the time I got to her there was red dirt and mud everywhere. And my gran-ma was immaculately clean. Her house was spotless... And so when I showed up, on her doorstep, full of Anahola red dirt and mud... I just think, what did my gran-ma think when she saw me, coming her way?

'But not once do I remember being turned away from her home, not once. What she would do is she would take me in the outside cement tub. And she would wash the mud off me. And then she'd take me in the inside bathtub and I remember my gran-ma is the only one who would scrub me clean.

'You know we were on our own as children: if we took a bath, we took a bath – if we didn't, we didn't. There was no hot water so most of the time we didn't until we were forced to. But my grandmother would scrub me clean, to get all the dirt out of my very long hair. And then... she'd sit me at her knee, and she'd patiently take every tangle out of my hair... And I'm crying cos it hurts and she's saying to me 'almost pau' – Hawaiian word for finished. 'Almost pau' – very gentle. 'Almost pau.' And some-times finishing would take an hour... I'm sitting at her knee for an hour. But she would be eventually pau, and I remember I'd stand up, and she'd take that comb and she'd go all the way down the back. And I remember as a little girl just feeling clean. And feeling pretty. And feeling like maybe somebody could love me today, maybe I'm OK today. That's what my gran-ma did for me. Just made me feel like I was OK.'

Mirena also thinks the boarding school she went to when she was 12 helped. 'I realised when I came here and I lived in the dorm, with all these different people, that families didn't have to be like this,' she says. The school's sense of community was important for her, and she remains working there today. It's also where she met her future husband, with whom she now has seven children and 15 grandchildren of her own. She says she recalls her grandmother often, particularly when thinking how she wants to be with her family.

'I remember on some of my darkest hours, raising these chil-dren in my life, thinking about her and knowing that I need to give as much as she gave to me. There is nothing that surpasses for me that example of love and caring. So I do my best to be that kind of gran-ma to my own.'

§ § §

It seems blindingly obvious that how we are cared for by our parents or primary caregivers is crucial, but the growing realisation of just how important love and affection are to children has only come about in the last century. Many of the studies that helped us to understand how childhood experiences can affect our adult selves hadn't been published back when Mirena and the rest of the Kauai cohort were born.

Some of what we know about the effect of parenting comes from watching animals. At Stanford University in the 1930s, in a series of experiments that would be unlikely to get through an ethics committee today, Harry Harlow separated baby rhesus monkeys from their mothers, and raised them in separate cages. He allowed the baby monkeys access to two models of a larger monkey: one made only of wire, but with a bottle of milk attached, and one with no milk attached but which was covered in a soft terry-towelling-type material. The young monkeys spent all their time on the soft model mother, craving the comfort, and only went to the wire one for food, before quickly returning to the towelled surrogate. This put into question all previous ideas about food and shelter being the primary drives for an infant, and suggested that the role of comfort might be much more important than was previously thought.

Bowlby was interested in what happened to children who were separated from their caregivers early on. One of his earliest studies was of 88 adolescent patients from his clinic in London. Half had been referred for stealing, and half had emotional troubles but had not shown delinquent behaviour. Bowlby noticed that the '44 thieves', as he called them, were much more likely than the control group to have lost a caregiver when they were young, which led him to think about how early experiences of loss can have profound effects.

Bowlby went on to write extensively about the importance of attachment and loss of attachment figures, influencing his colleague Mary Ainsworth to develop a way of measuring the quality of attachment between a caregiver and child, which is still used today. The 'strange situation', as it's called, involves observing a child's reaction to their caregiver leaving the room and later returning, and also their reaction to a stranger. Based on their reactions, their attachment can be classified in ways that can partly predict their later development. The most worrying classification, 'disorganised attachment', tends to be seen in children whose attachment figures have caused them harm, and has been linked to much poorer abilities to relate to others and regulate emotions in later life.

§§§

In the Kauai study, the children living in adverse circumstances largely remained in their homes, and some of them thrived regardless. But across the other side of the world, anyone in Europe old enough to watch TV in 1990 is likely to have a memory of the Romanian orphans. Images of children found in orphanages after the collapse of Nicolae Ceausescu's rule are deeply sad: bleak rooms, packed full of small children with big eyes, pulling themselves up on their cot bars to see the Western camera operators filming them. Under Ceausescu, abortion and contraception had been banned, leading to a massive rise in birth rates. Children without anyone to care for them had been left in institutions, to experience immense emotional deprivation and neglect. They had very little individualised care, no one to hug them or comfort them, no one to sing them to sleep. Their basic physical needs were met in terms of being given food and kept warm, but their basic emotional needs for affection and comfort were not. They learned not to even bother reaching out when adults were around.

The discovery of the conditions in the orphanages prompted a rush of compassion and charity initiatives to adopt the children. The UK Department of Health contacted a researcher at King's College London's Institute of Psychiatry, Psychology & Neuroscience, Michael Rutter, to ask him to measure what was going on.

'Like everyone else, I saw the media,' explains Rutter, sitting with me in his light and airy office at the Social Developmental and Genetic Psychiatry Centre in south London. 'But [the research] all started because the Department of Health contacted me, to say they didn't know what was going to happen to these kids, would it be possible to do a study, follow them through, and find out what were the policy and practice implications? ... So I said, let's have a go.'

For Rutter, this was a scientific opportunity as well as a practical one: 'This was a natural experiment.' All previous studies of children in care had involved groups of children who had entered institutions at a range of ages, meaning that variation in their behaviour and wellbeing might be related to things that had happened before they were in care. The Romanian orphans, though, had all been admitted within the first two weeks of life. 'It's a horrible thing to have happened,' says Rutter, 'but given that it did happen, one may as well learn as much as possible.'

Rutter's study assessed the children over time as they settled into new adoptive families. 'The findings were surprises all along the line,' he says. Prevailing wisdom at the time was that serious adversity in childhood led to a range of emotional and behavioural problems. Rutter's research found something different when the children were followed up: apart from a minority who had specific patterns of extreme social difficulties, such as autistic spectrum disorders, 'There was no increase in the ordinary emotional and behavioural problems,' he says. 'So that was one surprise.' Another surprise was that if the children were adopted out of care early enough – within six months – then they seemed to go on to develop well.

Rutter sees this resilience in the face of adversity as a dynamic process: 'Resilience initially was talked about as if it were a trait, and it's become clear that's quite the wrong way of looking at it,' he says. 'It's a process, it's not a thing.

'You can be resilient to some things and not others,' he explains. 'And you can be resilient in some circumstances and not others.' He acknowledges that 'children, or for that matter adults, who are resilient to some sorts of things are more likely to be resilient to others,' but he stresses that resilience is not a fixed trait.

Rutter offers a medical analogy: 'The way to protect children against infections is either to allow natural immunity to develop or to immunise.' Either way, children benefit from limited early exposure to pathogens. To prevent this from happening is, in the long term, harmful. Likewise, children need some stress in their lives, so they can learn to cope with it. 'Development involves both change and challenge and also continuity,' says Rutter. 'So to see the norm as stability is wrong.'

This suggests that there is something about the way that some children adapt to and cope with adverse circumstances that enables them to be emotionally resilient. It's not the stress itself that inevitably causes problems, although in the face of enormous adversity it would be much harder to remain resilient, but it's the interaction between the stress and the ways of coping that is really important. Maybe some ways of coping are more helpful than others, and maybe some protective factors mean that the stress gets managed better.

Rutter recalls a child he saw early on from the Romanian cohort who was really struggling with his behaviour and emotional wellbeing, but who has now gone on to develop in a seemingly resilient way. 'He has done very well,' says Rutter. 'Relationships at home are splendid, so there was a complete turnaround and it's difficult to know precisely why that happened, but the fact that it did happen reminds you that it's a mistake to write off situations as if they can't be changed.'

§§§

What if there are some children who need extra help, though, to boost them up to the same level of development as their more resilient peers? We still know very little about the mechanisms involved in resilience and how we can help them to be more effective. If we think of it as an adaptive process, how do our brains, our thought processes and our behaviours change to help us to cope with adverse early circumstances? Eamon McCrory, Professor of Developmental Neuroscience and Psychopathology at University College London, is investigating just this.

McCrory and his team are collecting a combination of brain images, cognitive assessments, DNA and perceptual data, from children who have been maltreated and allocated a social worker, and also from a control group who have not. The two groups have been painstakingly matched by age, pubertal development, IQ, socioeconomic status, ethnicity and sex. The researchers aim to follow their cohort for as long as funding allows, trying to unpick what would predict which of the children who have been maltreated will go on to develop difficulties and which will be resilient.

McCrory used to work clinically for the National Society for the Prevention of Cruelty to Children and he understands the clinical challenges that are involved with this population: 'Resources are very limited,' he explains, 'so if you have a hundred children referred to social services who experienced maltreatment, we know that the majority of them actually won't develop a mental health problem. But then a minority are at significantly elevated risk... At the moment, we have no reliable way of knowing which kid is which. So it seems sensible to try and move the focus back from the disorder to a much earlier stage in the process and characterise the risk profile... Only longitudinal designs can give us this information.'

McCrory's research is searching for reliable clues that a child will go on to develop difficulties, so that we can begin to know who to target to help. So far, McCrory has identified three main areas where there are likely to be differences: threat processing, brain structure, and autobiographical memory.

Studies of war veterans as well as maltreated children reveal that areas of the brain involved in processing threats, such as the amygdala, are more responsive both in the soldiers coming back from war and in children who have experienced early abuse. It makes sense that if you have been in danger a lot, then your brain may have adapted to be very sensitive to threat. 'Our main theoretical proposal at the moment is around a concept of latent vulnerability,' McCrory says, 'which is the idea that maltreatment... leads a number of biological and neurocognitive systems to adapt to a context characterised by early stress, threat and unpredictability, and adaptations to those systems may be adaptive and helpful in that context, but embed vulnerability in the longer term.'

The team are also scanning the children's brains to try to see whether difference in brain structure in maltreated children are stable over time or changeable. 'We know very little about malleability of brain structure over time,' explains McCrory. 'We know there are structural differences in the orbitofrontal cortex and the mediotemporal lobe, for example, which are quite robust, but we've no idea whether they are static or whether they may shift over time, at least in certain children.'

The third area the team think is important is autobiographical memory. The brain system involved in thinking about and processing memories of personal history might also be shaped by early traumatic experiences in a way that is adaptive in the short term but unhelpful in the longer term.

'Autobiographical memory is the process whereby you record and encode your own experiences and make sense of [them],' explains McCrory. 'We know that individuals who have depression and PTSD [post-traumatic stress disorder] have... an over-general autobiographical memory pattern, where they lack specificity in their recall of past experience... We also know that kids who have experienced maltreatment can show higher levels of this over-general memory pattern. And longitudinal studies have shown that a pattern of over-general memory can act as a risk factor for future disorder.

'One hypothesis is that the over-general memory limits an individual's ability to effectively assimilate and negotiate future experiences, because we draw on our past experiences to be able to predict the contingencies and likelihood of events in the future, and use that knowledge to negotiate those experiences well. So... over-general memory might limit one's ability to negotiate future stressors.'

It makes sense that if horrible things have happened to you in the past, you will want to avoid thinking about and remembering them, which might lead to a tendency to have a memory that's light on detail. McCrory's team are finding reliable associations between over-general memory patterns and childhood maltreatment.

Back to Mirena in Hawaii, and she finds it hard to know whether her memory has been affected by her early experiences: 'from a personal perspective I wouldn't know,' she says. 'We don't know what we don't remember.' The memories she does have of her family growing up are mixed. In our conversations, she often describes them fondly: her father as 'a brilliant man' who 'read all the time' and was 'just kind of ordinary except when he was drunk', and her mother as 'a beautiful Hawaiian woman who had a beautiful voice, who did her best'. Alongside these descriptions are darker memories, of coming home to arguments in the kitchen, or worse: 'I saw my mother try to kill my father on several occasions, cos daddy was drunk and mom was mad. And I was usually the one that would try to stop them.' While we talk, Mirena sometimes becomes tearful, remembering difficult times, and other times speaks with passion about the importance of protecting other children.

§ § §

In an ideal world, we wouldn't have to work out how to best to help children who have been abused or neglected; we would instead be able to remove those risks. Admitting that we don't live in that ideal world, and trying to understand what we can do to prevent the negative effects of childhood adversity and to boost individual resilience, is perhaps the next best thing.

Everyone I interviewed for this piece had a sense of optimism. 'That's the psychological perspective, right?' says Lali McCubbin. 'We want to believe that people can turn their lives around.'

McCrory certainly does: 'I think it's hopeful to see that recovery is possible and that these [brain systems] are systems characterised by plasticity, and so the questions are then about how do you promote that, are there developmental periods where that is more possible, and how much can we enhance plasticity over those periods?'

The concept of childhood resilience is complex. McCubbin recalls a conversation she had with her father and Emmy Werner about the use of the term, discussing whether they would have called it resilience if they had known then how much it would take off. 'And they weren't sure if they would, and I liked that... because it's really about adaptation... A lot of people miss that take-home message, and that 'Oh, the individual wasn't resilient', it kind of blames the individual rather than looking at their context. What may be resilient for you may not be the same for somebody else.'

The idea of resilience as an adaptive process rather than an individual trait opens up the potential for other people to be involved in that process. McCubbin sees the importance of relationships as being wider than only protective relationships with people, and she and her team have created a new measure of 'relational wellbeing' to try to capture this. 'We think of relationship as with a person,' she says. 'But what we really found was that it was relationship with the land, relationship with nature, relationship with God, relationship with ancestors, relationship with culture.'

McCubbin's team have just finished pilot interviews with eight of the original cohort, now in their 60s. She weaves in the Hawaiian idea of aloha as she describes the research. 'There's a tourist version of aloha,' she explains, talking about a word that is variably translated as 'love and compassion', 'mercy' and 'connectedness' or 'being part of all and all being part of me'.

'Aloha means hello and goodbye, but actually aloha means 'breath of life',' McCubbin continues. 'That was one of the things in our interviews, we were collecting their mana'o, their life's breath... We got chicken skin when you hear it that way, just that sense of aloha and that sense of how we're all connected.'

Mirena is clear about the importance of human connection, and so is the research, although we have a way to go before what we are learning about how to best care for children who have survived childhood maltreatment is clearly understood and communicated to all those working with children. For Mirena, the vital thing is still 'that there's somebody they know cares about them. Just one person, it can make all the difference.'

This story was first published on 21 June 2016
by Wellcome on mosaicscience.com

* Her name has been changed.

What tail-chasing dogs reveal about humans

■ Shayla Love

Curiously, and perhaps eagerly, I am looking at a bull terrier named Sputnik, searching for a resemblance.

He's a stocky three-year-old, mostly slate grey, with a white stripe on his head and a pink splotch on his elongated, bull-terrier nose. So far, our only similarity is we're both waiting in a trailer that's serving as his examination room at Tufts University's veterinary school in North Grafton, Massachusetts.

Sputnik has canine compulsive disorder (CCD) and is at Tufts for a checkup with Nicholas Dodman, a veterinarian who has been studying CCD for over two decades. I'm shadowing this visit to learn about Dodman's work and, selfishly, to learn something about myself; I was diagnosed with obsessive–compulsive disorder (OCD) a few months ago.

When Dodman first started seeing these dogs, he realised he had been handed a potentially ideal animal model to study human OCD. But in 20 years of studying dogs, discovering genes that may be involved and new neural pathways, one problem has continually clouded his research: the debate over whether CCD can truly be compared to human OCD. 'When it comes to problems of the mind, people have a mental block,' he says. 'The mind is thought of as uniquely human.'

I try to see into Sputnik's eyes. He stays close to his owner, Dan Schmuck, giving me an occasional glance. Sputnik was a tail chaser, and would spin for hours on end. At the moment, he is completely still. Like me, it seems he keeps that kind of behaviour away from the public eye.

Two years ago, after rescuing Sputnik from a shelter, Schmuck went on a business trip. His mother called him to say that their new dog had started chasing his tail, and she couldn't get him to stop. At first, Schmuck and his wife thought it was funny. They took a video of their young puppy spinning, and you could hear them laughing in the background. But soon, the humour faded.

'It was as if I didn't exist,' Schmuck explains. 'His head will lurch all the way against his shoulder, and stare at his tail, as if his nemesis is staring right back at him. He will slowly start working up to chasing, and it will get faster and faster until his head is hitting whatever wall he bangs into. Even though he's getting hit so hard that you think he's getting a concussion, he will keep doing it until his teeth and his tail start hitting the wall and he's shooting blood all over the place.'

Schmuck had to take time off work to stay home and physically restrain him. He held Sputnik's head in one hand, and wrapped his arm around his rear until he calmed down or fell asleep. 'Then he would wake up and I could just feel him... you could feel his head start lurching like he's thinking about it. It was an unsustainable situation.'

Schmuck drove from his home in Baltimore to see Dodman in Massachusetts. Dodman had seen this kind of spinning many times before. Tail chasing is a common compulsive behaviour for dogs, and for bull terriers specifically. Particular breeds have particular behaviours they exhibit with CCD. Bull terriers spin, Dobermans lick their limbs and suck their flanks, Labradors hold objects or chew rocks, and King Charles spaniels snap at imaginary flies.

Like human OCD, which is commonly focused on washing, hoarding, counting or checking, canine compulsions fit into neat categories. And while these behaviours may sound trivial, they are performed to the extreme. They take over eating, sleeping and all basic functioning. In some cases, they can be fatal.

Sputnik steps out from the safety of Dodman's shadow, timidly eating treats. His tail hangs innocently between his legs, and I'm having a hard time imagining him whirling, manic, out of control in a pool of his own blood. 'Over the last two years, he has gone from being a dog that was going to have to be put down, to being a normal dog,' says Schmuck. 'He maybe looks at his tail once a day. It's a miracle.'

Sputnik is on Prozac now, along with a few other medications to temper his behaviour. Could Sputnik be like me, a person with OCD? Was he thinking about his tail right now, somewhere in the recesses of his doggy mind?

'You can't access an animal's thoughts, so the purists call it only canine compulsive disorder, not obsessive–compulsive disorder,' Dodman says. 'But it looks for all the world, when Dan is holding Sputnik back, that he's constantly thinking about it. That's an obsession.'

I was seven, maybe six, when I realised that my hands were dirty. I could wash them, and then they would be clean. But if I touched something – the banister, my clothes, the sofa, a door handle – they would be dirty again. That was easy enough to fix, just wash again. Until I touched something else, and the process would repeat.

At home, it wasn't an issue. I could wash my hands whenever I felt my hands weren't clean. But at school, snack time, field trips – there wasn't always a place I could wash my hands. Or, the catch-22, the public bathroom at school or the cinema wasn't clean enough. I couldn't really be sure that my hands were clean.

I devised a whole bunch of techniques to maintain my hands' cleanliness. When I went out for lunch on the weekends with my parents, I would wash before we left, close my fists tightly, and hide them in my sleeves until we got to the restaurant. Or I wouldn't eat the parts of my food that I had held with my hands. I left the bottom halves of French fries and discarded the corners of sandwiches.

I invented a game that I played with my friends at the movies: bobbing for popcorn with our tongues. When the lights dimmed, they returned to eating their popcorn normally; I continued to bob.

This went on for a couple years, until the fourth grade. I still remember the first food I ate without washing my hands. They were fruit gummy snacks handed out at an after-school programme I went to. I opened the plastic bag, reached in with my fingers and put them straight into my mouth. I may have even licked the tip of my index finger a bit. I didn't know why I was suddenly different, but I felt exhilarated. I was free, normal. I didn't know what had kept me chained those past few years, but I was glad it was over. I had seconds that day, in my favourite flavour, grape.

In 1989, a popular science book called *The Boy Who Couldn't Stop Washing* was published. Its author, Judith Rapoport, then chief of the Child Psychiatry Branch at the National Institute of Mental Health, had studied and treated all kinds of neuropsychiatric illness, but a fascination with OCD grabbed hold of her. People with OCD had to participate in detailed rituals and compulsions to assuage strange beliefs: that they had just killed someone, that everything was contaminated, that they had sinned in some way, that things had to be just right.

Before her book came out, OCD was thought to be rare. We now know it affects 1–3 per cent of the population. Rapoport's book was one of OCD's first steps into the spotlight – she went on *Oprah* and *Larry King Live*. Millions of people began to understand something about their own odd behaviours, or those of their friends and family members. Soon, Rapoport started to get letters and phone calls – including some with questions she hadn't expected.

'A whole bunch of them talked about their dogs,' she says.

People wrote that their dogs did these behaviours too, especially excessive washing. Did she think they had OCD? It was an interesting idea. 'If people ask one weird question, you shrug it off,' she says. 'But if 20 of them ask it, you pay attention.'

A dog owner herself, she went to her vet to ask about acral lick, a common CCD behaviour when a dog licks or sucks at its paw or leg until the fur and flesh are worn through, leading to infection, amputation and sometimes death. Her vet told her yes, acral lick was a huge problem with no good treatment options, and his dog suffered from it. She asked if he would be willing to try medication – the same medication given to people with OCD, which increases levels of the brain chemical serotonin.

'We put his dog on a dose that we guessed at, given the weight of dogs and the weight of people, and the dog had a remarkable response,' she said. 'You could say this all started when I cured my vet's dog.'

Encouraged, Rapoport designed a double-blind controlled study. Dogs with acral lick received one of two drugs for OCD that targeted serotonin, or a placebo, or an antidepressant that worked for depression but not OCD and wouldn't alter serotonin levels. The results were 'dramatic': the only group that improved was the group that got the serotonin drugs.

Still, Rapoport took her findings with a grain of salt. As a psychiatrist, she says she usually needs to know what her patients think of their compulsions to give them a true OCD diagnosis: 'Patients with OCD have insight and they say, 'Look, this is very embarrassing, I think this is crazy what I'm doing, but I can't stop," she explains. 'Well, you can't get that sort of information from animals, so animal models are often very limited for psychiatry.'

After publishing her findings, she moved back to human patients. But her work caught the attention of a veterinary anaesthesiologist with an interest in behaviour: Nicholas Dodman.

§ § §

After an injection of morphine, a small black horse named Knightly Night began to repetitively paw at the ground.

It was the 1980s, and Dodman had noticed that by using different drugs he could change the way animals behaved. In horses, he could 'turn on' certain repetitive disorders that in the equestrian world were called stall walking or cribbing. Together with Louis Shuster, a professor of biochemistry and pharmacology at Tufts

School of Medicine, he asked the question that would launch his career in animal behaviour: if they could turn a behaviour on, could they also turn it off?

Dodman and Shuster gathered horses with severe problems, who would pace and bite at guardrails, and gave them narcotic antagonists – the opposite of morphine. They all stopped.

'This behaviour has been going on for two thousand years of horses in captivity and nobody knew why,' says Dodman. 'We were able to show that somehow you can turn it on and turn it off; that it was caused by some neurotransmitter imbalance.'

Dodman and Shuster's initial hypothesis was that the drugs were blocking the brain's natural opiates, stopping the behaviour because the animal no longer felt good while doing it. In further experiments, that theory didn't hold up. So then they theorised that the on–off switch they had discovered actually lay in the drugs' effects on NMDA receptors, which interact with a brain chemical called glutamate. Dodman and Shuster thought maybe cutting the line to glutamate was somehow stopping the behaviour.

To test their theory, Shuster went to a local store and bought a bottle of Delsym, a cough suppressant containing dextromethorphan, which also blocks NMDA. They fed it to a pony named Cinnamon Bun, a severe crib biter.

'He glugged it down and it stopped his cribbing behaviour,' Shuster says. 'The cough medicine made his lips all pink, he looked rather strange, but it worked.'

Dodman opened an animal behaviour clinic at Tufts to expand his research to other kinds of animals, and the patients began to crawl in. He saw all kinds of pets: more horses, cats and birds, but he began to narrow his focus to dogs.

'The reason canine genetics are so cool is one word,' Elaine Ostrander tells me. 'Breeds.'

Ostrander is the chief of the Cancer Genetics and Comparative Genomics Branch at the National Human Genome Research Institute, and has been working in dog genetics for 25 years. Her lab develops dog genome databases to look for genes that could be important for animal health or translate to humans. She says they've explored everything from infectious disease to cancer, including diabetes, kidney failure, retinitis pigmentosa and gout.

'If you want to understand the genetic underpinning of a complex disease, we know there's lots of genes involved,' she says. 'In human populations, there are dozens of genes that contribute. Every family is a little bit different. Some genes seem hereditary, some seem not to be, it's a very complex mosaic. In dogs, you simplify that mosaic.'

Within breeds, dogs are genetically very similar. But also between breeds of related dogs, Ostrander can see commonalities. By looking for disease genes in sick dogs in closely related breeds, she can exclude false positives; if four similar breeds with a disease all carry the same gene, one that unaffected dogs don't have, she knows she's got a strong candidate.

In 1994, Dodman teamed up with Alice Moon-Fanelli, an animal-behaviour geneticist, to help him begin to explore the genetics of his dog patients. Ostrander had the genetic data and Moon-Fanelli started to gather phenotypes, the expression of those genes: including the details of each dog's behaviour, along with its breed, pedigree, age of onset, and anything else that might be useful.

Moon-Fanelli says when they started their project, the idea of CCD – not even as a model, just as a standalone disorder – wasn't widely accepted. Animal repetitive behaviours were considered 'stereotypies' – mindless actions that were a result of poor environment or boredom, like tigers pacing in a cage at the zoo. 'I came into it asking, 'Why is this any different?'' she says. 'And looking at almost 400 bull terriers over the years, and all the Dobermans and cats, it became clear that it wasn't because of a suboptimal environment. These animals were pets, they had great lives in wonderful homes.'

The dogs' symptoms usually started around puberty, as is often the case in people. Compulsive behaviours ran in family lines, just like people. And just as human psychology had to realise that human OCD wasn't a result of upbringing, animal medicine was doing the same.

'One living species to another, you know that they are obsessing and that they are possessed by demons that they cannot control,' says Moon-Fanelli. 'And it's the same thing with people. It's just that people speak the same language so they can tell us what they're thinking. We have to develop our interpretation, and try to be objective, based on what we see in its behaviour.'

Pamela Perry, a behaviourist at Cornell University's College of Veterinary Medicine, isn't involved in Dodman's work. She treats animals that have a variety of behaviour problems, and says that while stereotypies and compulsions can often overlap, she does recognise a distinction. She agrees that we don't know for sure if animals are obsessing. But she has seen dogs who don't just compulsively chase light and shadows: they even get up before dawn and wait for the sun to cast shadows that they can then chase. Another client had bought a new washing machine, and their dog would wait for it to turn on and spin along with the whole cycle. As soon as it stopped, he would stop.

'If they're waiting and anticipating, personally I think we can presume that they're obsessing to some degree,' Perry says.

§ § §

Despite having one of the classic presentations of OCD, hand-washing, it never occurred to me or anyone who knew me that I had OCD. I've always been anxious, enough so that I sought out talk therapy several years ago and considered myself a person with generalised anxiety disorder.

My anxieties were always around a common theme: cleanliness, disease, health, germs. I also have a phobia of vomiting, which I obsess about daily. I think about throwing up probably 10–14 hours out of the day, and actively avoid situations that I think will cause me to feel sick or be sick. I thought this was what anxiety was – to be concerned about a specific set of things.

In psychotherapy, we discussed the 'reasons' for my anxieties. My parents are scientists and I learned about germs at a very early age. My father was also concerned with germs and cleanliness, food poisoning and food safety. I liked to be in control, and throwing up was a total loss of control, a window into vulnerability. These sessions made me feel better, and I do think my obsessions diminished a little bit as a result. I felt that knowing their origins and roots would help me manage them, help me talk them down.

Looking back, I think my inability to recognise that I had OCD was rooted in the belief that, as a human, I had control over my behaviours and thoughts. Or that, if I didn't, it was because of deeper human thoughts – if I could just uncover them, I would regain control.

But in situations where I was challenged, I quickly saw how little control I had. When I accidentally ate something that had gone rotten, I descended into total and utter panic for days. When my boyfriend got the stomach flu, I fled our apartment, staying in a hotel for three nights. When I got home, I bought hospital-grade cleaner and bleached and cleaned our home. I didn't feel safe for weeks; every day I thought obsessively about the germs that were still present, waiting to infect me.

Some of this may sound understandable, albeit a little over the top. But like the dogs, or others with OCD, it's the amount of disruption of daily life that determines a disorder. I was spending hours upon hours thinking, cleaning, worrying, obsessing.

Throughout all this, I still didn't think I had OCD, and it wasn't until my physical compulsions returned, at the end of last year, that I finally thought it might be more than anxiety. I travelled for work from May to December of last year, and dealt with some very minor health problems. Whether through those triggers, stress, isolation, or some combination, the compulsions began to creep back up on me.

Like my childhood handwashing, I don't know when exactly they began. Now that I'm doing them, it's as if they're permanent fixtures, though I can look back six months ago and recognise that I was free of them. My current rituals mostly centre on eating, swallowing and food safety.

Right now, when I eat, I have to eat completely alone, and if someone else is in the room, I cannot eat. I just... can't. When I do eat, if the food needs to be heated in the microwave, I have to set it to an odd number, usually 4 minutes and 37 seconds. Then, I have to stop the microwave at 37 seconds, though 27 or 17 are acceptable also. (While doing this, I recall with an odd fondness that in my childhood the number on the microwave always had to be 29. It's like remembering an imaginary friend.) As I'm eating, each time I swallow, I have to lightly touch or grasp the tip of my nose and look up to the far-left corner of my field of vision.

In addition to the eating rituals, my other obsessions got much worse. I started to throw away food compulsively, food I knew – logically – was okay to eat. But fears of contamination take hold: what if it's gone bad? What if the refrigerator isn't cold enough? What if this bag of frozen berries was in a place in the freezer where the air didn't circulate properly?

The fear and obsession with vomiting became extreme. Anything that looked or reminded me of throw-up could cause panic. Spilled coffee, splatters of water, soup, anything creamy, the words 'chunk' or 'throw'. Even writing this single paragraph, using these words, has taken me hours. Each time I write 'throw-up' or 'vomit', I have to take a break.

Last year, when I was travelling, I wasn't having any kind of treatment, not even talk therapy. When I got back to New York in January, I enrolled in a clinical trial for people with anxiety at Columbia University. While being evaluated, my interviewers casually asked, 'Do you have any phobias or anxieties that aren't related to your OCD?'

The question stopped me in my tracks. I have OCD?

§ § §

We don't know exactly what goes wrong in the brain to cause OCD. We do know that a group of drugs known as SSRIs (like Prozac), which increase serotonin levels, seem to help – but only for some people. About half of people see a response to the SSRIs, and a 'successful' response can mean as little as a 35 per cent reduction of symptoms. As a recent review of OCD treatments said: 'This means that even treatment-responsive patients may continue to have levels of symptoms in the mild-to-moderate range and spend hours daily preoccupied with their obsessions and compulsions.'

Whether you believe in a dog model or not, one thing is becoming clearer in human OCD research: serotonin is not the complete story.

As Dodman had noticed, glutamate seems to be important. Recent neuroimaging of people with OCD has shown higher blood flow and activation in the cortico-striato-thalamocortical circuits, a network that loops from the deep centres of the brain to the prefrontal cortex. This area is dominated by glutamate pathways that are believed to generate controlled movement and thought, and to modulate behavioural routines. Some OCD researchers now hypothesise that SSRIs work not because of serotonin, but because they stop the release of glutamate. Further work, testing the cerebral spinal fluid levels of people with OCD, found that they had significantly higher levels of glutamate.

Still, knowing that glutamate plays a role, in dogs or in people, doesn't help discover the genes that cause this disorder in humans, which is where an accurate animal model could be extremely helpful.

'The problem with the large number of behavioural disorders is that we don't really have a good clue as to what the underlying molecular change is,' says Ed Ginns, a neurologist and geneticist who works with Dodman. 'If we can at least get that, we're confident that, with further molecular and clinical studies, pathways and even potential targets for treatment can be identified.'

When he first met Dodman, Ginns had been studying diseases like bipolar disorder and depression in genetically closed populations like the Amish. For him, it was never an issue that Dodman's sample was built of Dobermans and terriers. He says it was compelling because, like the conditions of those Amish – and unlike other animal models – the dogs' disorder had arisen naturally.

'These are not artificial constructs,' he says. 'These are patients walking into his office with a real behavioural issue. It doesn't rely on us guessing what we think might be the gene or the change in a mouse that might model a disease.'

Ginns and Dodman's first collaboration was a genome-wide analysis, comparing 92 flank- and blanket-sucking Doberman pinschers with 68 control Dobermans. They got one strong statistical hit in what Dodman calls a 'genetic oasis' – there was only one large gene there for them to look at, called neural cadherin or CDH2. In the brain, CDH2 is involved in the development of glutamate receptors.

'It was a great gene,' Dodman says. 'Everyone basically took a big breath and stepped back. This was the first behavioural gene that had ever had anything to do with OCD, and one of the few behavioural genes that have been discovered.'

The next step was to look for CDH2 in people. Dodman and Ginns took their research to the National Institutes of Health, and a group there analysed data they had from people with OCD. The results were inconclusive. They found a suggestion that some CDH2 variants might be associated with Tourette syndrome, but that picture was fuzzy as well.

'We didn't find something earth-shattering,' says Jens Wendland, a physician and psychiatrist who co-authored the study. 'But to be fair, we know now that the cohort would have needed to be much, much larger, at least an order of magnitude larger, to be properly powered to do that. And we tried the best we could with the means we had available at the time.' Wendland thinks that sequencing has advanced enough that it's more beneficial to study humans than to redo any dog studies. He is sceptical whether we can ever really be sure that a dog's symptoms can correspond to humans'. 'We will never really know that for sure, so you could argue why should we take this leap of faith in the first place?' he says.

'I would much rather start to work on the biology of genes identified from human studies, however challenging that may be. As opposed to starting with a gene mapped to a behaviour in nonhumans where I can never be certain that this is really affective of the condition that I want to treat.'

In 2008, Dodman decided to take the initiative and move his theories to a clinical setting. For many years, he had been discussing his work with Michael Jenike, founder of the Obsessive Compulsive Disorder Institute at McLean Hospital in Belmont, Massachusetts. Jenike enjoyed his talks with Dodman but wasn't convinced. Like Judith Rapoport, he says that the trouble with dogs and birds and mice is that unless he can talk to them, he can't properly diagnose OCD.

Still, he was willing to try giving some of his patients memantine, a glutamate-targeting drug normally used to treat Alzheimer's, which Dodman had started giving to dogs with severe CCD.

In a group of 44 patients, everyone got a drug to increase serotonin levels, but half were given memantine as well – and it worked. For those who also got the glutamate drug, symptoms reduced by 27 per cent on average, compared to 16.5 per cent for the others. It isn't perfect, but Jenike continues to use this combination of drugs with patients who aren't responding well to SSRIs.

Dodman and Shuster had already patented the combination of drugs as a treatment for OCD, but their tech transfer office at Tufts couldn't get any pharmaceutical companies interested. However, subsequent research has supported the idea that both serotonin and glutamate pathways need to be addressed when treating OCD.

In brain imaging of his compulsive Dobermans, Dodman found that they had structural abnormalities associated with OCD in humans. In February 2016, a group led by Dan Stein, head of the Department of Psychiatry and Mental Health at the University of Cape Town, published the results of a re-examination of the CDH2 gene in humans. Their sample was made of 234 people with OCD and 180 healthy controls, and their findings were more conclusive than the previous study: they found two differences in the CDH2 gene that seemed to be correlated with OCD, though Stein says more work is needed to fully understand the connection.

Dodman's latest work, published in 2016, compares dogs that have severe and mild cases of CCD. He found two areas of interest in the genome. The first has a human counterpart that is associated with an increased risk of schizophrenia, and the other harbours serotonin receptor genes.

From these most recent findings and their connection with serotonin, Dodman has a new theory. He thinks that the CDH2 gene, which involves glutamate, is required for a dog to be genetically predisposed to CCD in the first place. A human might have a different predisposition gene, but Dodman guesses it involves glutamate too. Serotonin genes, he thinks, are modifiers that control to what degree a dog has CCD (or a person has OCD). He now hopes that someone will look for similar modifier genes in humans, or expand standard OCD treatment to include both serotonin and glutamate pathways.

Dodman still thinks that any hesitation to accept research based on a dog model of human OCD doesn't lie in specific doubts about the validity of the model, but in a greater philosophical problem: the difficulty in accepting that our minds might be closer than we want to believe to the minds of dogs.

'It really helps to be a veterinarian,' he says. 'Because one of the things people say when you're a vet is, they say: 'It must be so difficult because you have to learn all these differences between the various species.' The answer is, actually you don't. What you learn to do is appreciate the sameness.'

§ § §

As my evaluations at Columbia have continued, I've come face to face with another kind of sameness. It's unsettling, almost spooky – as if someone has entered the most private, most anxious parts of my brain, written down my personal thoughts, and then put them on a questionnaire to be read out loud by a bored psychiatrist. Every worry, every obsession, the things I have to do in secret are standardised and stereotypical enough to be on these general evaluation forms.

Soon after I was officially enrolled in the clinical trial, I began cognitive behavioural therapy (CBT). It is the exact opposite of my past therapy sessions. I loved my psychodynamic therapist. We would meet once a week in her pleasant office, with bookcases and Asian artwork hanging on the wall. She was achingly smart and well read, and as we talked about my feelings and childhood, we often discussed theories and psychoanalytic texts. It felt like an intellectual exchange, something human of the highest order. Deconstructing, understanding, breaking down symbols and metaphor into meaning.

At CBT, I am meeting twice a week on a hard plastic chair in a tiny office with no windows. But we are not here for my feelings; actually the opposite. I am here to provoke my anxieties, confront them and, hopefully, neutralise them. I'm beginning exposure therapy and response prevention. We will not be discussing my dreams. CBT feels as close to dog training as any human activity can get.

We meticulously go through all my phobias, obsessions and rituals and rate them on a hierarchy. Now my job is to attempt to cease the rituals, and expose myself to increasingly upsetting stimuli for an hour a day, and 90 minutes twice a week in session.

After my first exposure session, I was stunned the whole subway ride home. It was upsetting, but more because it was the first time I realised how sick I was. I had looked at photos of spilled orange juice, something I haven't been able to do because it reminds me of throw-up. First I looked at a cartoon, then more and more difficult photos (difficulty being based on how similar the colour and texture were to actual vomit).

I was so distressed by these photos I could barely glance at them before breaking out into my touching-and-looking swallowing ritual. There was a part of me that thought: this is ridiculous. It felt as if I was being tortured, but I was simply seeing a cup of juice overturned. How could this be making me so upset? I knew it was a cup of juice, and yet when I looked at it, the obsessions flowed in, thinking about being sick, thinking about if the image would make me sick, imagery of me or others getting sick, on and on.

Alice Moon-Fanelli said that the spinning dogs seemed like they had been taken over by demons. These were my demons. I didn't expect them to look like spilled juice.

I picked up a copy of Rapoport's book, now over 25 years old, and the same Twilight Zone-esque feelings from my evaluation interviews came back. The young boy who 'couldn't stop washing' also hid his fists in his sleeves and had a complicated swallowing ritual that involved touching and blinking. If the name had been removed, I could have easily believed someone wrote it about me.

In her book, Rapoport wrote that she was amazed at the sameness of OCD behaviours. Though she remains unconvinced how much animal models will ultimately help, she did say that psychiatrists could learn from the work of ethologists, who study inborn behaviour patterns in animals. Rapoport's collie dog turned in circles – not compulsively, but before it lay down to sleep. In the dog's ancestors, that behaviour was conserved to trample down grass or ward off hiding snakes or insects. But in her suburban home, it remained.

'The highly selective nature of OCD behaviours is just as remarkable,' she wrote. 'Washing, grooming, hoarding – any theory of this disease must account for the incredible selectivity of these behaviours, which could be action patterns from an ethologist's field book. As psychiatrists, we need to be field observers much more often than we are.'

Forcing myself to see myself as an animal, one that could be overtaken by innate behaviours, helps me make sense of what it means to have OCD. The endless thoughts of vomiting and cleanliness, and my eating rituals, are the same as a dog spinning round and round. The spinning even provided me an apt visual metaphor to latch on to. When the thoughts or fears begin, it is like a whirlpool: swirling, spinning, gaining momentum and sucking all logic or reason into the bottom, out of sight. As a human, I am enraged when I can't stop my obsessive thoughts. As an animal, it makes sense that I can't talk my way out of compulsions or fears. It is easy to accept that a dog's broken neural circuitry was causing their tail chasing. I'm working on lending myself that kind of compassion.

§ § §

The day after Sputnik, I meet Bella, another bull terrier who used to spin but after treatment has stopped almost completely. Bella's owner, Linda Rowe-Varone, has a similar tale to Sputnik's owners: one day her sweet puppy started to spin, and nothing she could do could stop her. Like Dan Schmuck, she tells me she almost reached breaking point.

'There was a time I really thought she was spinning so much that I could not keep her in my home anymore,' she says with tears in her eyes. 'And Dr Dodman just kept telling me, 'Just wait it out, wait it out, you've got to give her a little more time.' And I'm really glad I did.'

Bella is active, playing and running around the examination room. Rowe-Varone tells us that Bella is also obsessed with balls, and she has to limit what kinds of toys she is exposed to. I wonder where the vets draw the line? All dogs have a favourite toy, one they love to play with. When can they call it an obsession? (When did I go from being a person who liked to be clean, to becoming obsessed with cleanliness?) Rowe-Varone says that she has to keep balls hidden in the garage, and if Bella sees them, she will sit outside the garage door for hours.

The debate about whether dogs can truly obsess doesn't enter this room. The consensus here is that Bella knew her balls were in the garage and couldn't get them off her mind. I'm struck by how accepting dogs as an animal model for human OCD required two shifts in thinking: not only did we have to become more animal, but we had to grant them a bit more humanity as well.

Dodman remembers a dog that was obsessed with water. It lived mostly in New York City but when it went to the owner's Hamptons home it would jump into the pool and do laps for seven hours a day, whining in anxiety the whole time.

Stephanie Borns-Weil, who took over from Dodman as programme director at the Animal Behaviour Clinic at Tufts last year, has seen a golden retriever obsessed with water too: it would get into the bathtub with the kids, or stand in puddles on its walks and refuse to move. Another dog she saw a few months ago would go into a lake by its house, take out five rocks and put them by a tree. If the owner removed them, it would return to the lake and get the rocks back.

One Doberman needed to cover her food before she ate. When her owner fed her, she put some paper towels next to her food. The dog would take a paper towel very delicately in her mouth, cover the food, and then uncover the food to eat it. If she couldn't perform this ritual she wouldn't eat. Dodman remembers another odd eating ritual: a dog who would take individual pellets of food and place them in the button compressions in the couch cushion in the next room. Only when he had put seven pellets into seven button holes could he eat the rest of the food out of his bowl.

I look back at Bella, who has bored of our chatting and is resting underneath a desk. The eating rituals that Dodman and Borns-Weil are describing are hitting a bit close to home, and maybe that's why, for the first time, I do recognise a bit of myself in Bella. There is a big container of balls above her head, which we hurriedly covered when we came in. Is she thinking about them now, just like I'm thinking about my own obsessions?

When I leave Tufts that day, my boyfriend, Zach, picks me up and we head back to our hotel, tired and hungry. I don't have a driver's licence, and I needed Zach to drive me the three hours to rural Massachusetts. The drive up from Brooklyn was fraught with anxiety. Being in the car is one of my triggers. Obsessions about car sickness began weeks ago, when I first booked the rental. The whole drive I was tense and white-knuckled.

I know that I need to eat the moment we get back to our room. Zach jumps on the bed, settling in with a book. Nervously, I ask him: can you leave, so I can eat? The ritual needs to happen alone. He's frustrated. Now? He will have to go sit in the lobby. Fine. Annoyed, he grabs his coat and heads towards the door.

Now I'm upset. Upset I have to do the ritual, upset he can't be more understanding, mad at myself for having so little control. 'You think I want to do this?' I yell defensively. 'I just need a little time.'

'Whatever,' he says, slamming the door.

I realise I've made the same plea Dodman made to Rowe-Varone when she was about to give up on Bella: 'You've got to give her a little more time.'

I'm in tears as I set up for lunch. Three napkins, an odd number, and I warm up the soup stopping on 37. I eat, with the ritualised looking and touching, trying to not think about how my soup kind of looks like throw-up.

'It's not that they're dogs and it's not that we're humans,' Ginns said to me. 'It's that both groups are suffering from this same clinical presentation that disrupts their development and lives. And it's that description that for me defines compulsive behaviour.'

I take a deep breath. Just wait it out, wait it out, you've got to give yourself a little more time.

This story was first published on 25 June 2017
by Wellcome on mosaicscience.com

A central nervous cure for arthritis

■ Gaia Vince

When Maria Vrind, a former gymnast from Volendam in the Netherlands, found that the only way she could put her socks on in the morning was to lie on her back with her feet in the air, she had to accept that things had reached a crisis point. 'I had become so stiff I couldn't stand up,' she says. 'It was a great shock because I'm such an active person.'

It was 1993. Vrind was in her late 40s and working two jobs, athletics coach and a carer for disabled people, but her condition now began taking over her life. 'I had to stop my jobs and look for another one as I became increasingly disabled myself.' By the time she was diagnosed, seven years later, she was in severe pain and couldn't walk any more. Her knees, ankles, wrists, elbows and shoulder joints were hot and inflamed. It was rheumatoid arthritis, a common but incurable autoimmune disorder in which the body attacks its own cells, in this case the lining of the joints, producing chronic inflammation and bone deformity.

Waiting rooms outside rheumatoid arthritis clinics used to be full of people in wheelchairs. That doesn't happen as much now because of a new wave of drugs called biopharmaceuticals – such as highly targeted, genetically engineered proteins – which can really help. Not everyone feels better, however: even in countries with the best healthcare, at least 50 per cent of patients continue to suffer symptoms.

Like many patients, Vrind was given several different medications, including painkillers, a cancer drug called methotrexate to dampen her entire immune system, and biopharmaceuticals to block the production of specific inflammatory proteins. The drugs did their job well enough – at least, they did until one day in 2011, when they stopped working.

'I was on holiday with my family and my arthritis suddenly became terrible and I couldn't walk – my daughter-in-law had to wash me.' Vrind was rushed to hospital, where she was hooked up to an intravenous drip and given another cancer drug, one that targeted her white blood cells. 'It helped,' she admits, but she was nervous about relying on such a drug long-term.

Luckily, she would not have to. As she was resigning herself to a life of disability and monthly chemotherapy, a new treatment was being developed that would profoundly challenge our understanding of how the brain and body interact to control the immune system. It would open up a whole new approach to treating rheumatoid arthritis and other autoimmune diseases, using the nervous system to modify inflammation. It would even lead to research into how we might use our minds to stave off disease.

And, like many good ideas, it came from an unexpected source.

§ § §

Kevin Tracey, a neurosurgeon based in New York, is a man haunted by personal events – a man with a mission. 'My mother died from a brain tumour when I was five years old. It was very sudden and unexpected,' he says. 'And I learned from that experience that the brain – nerves – are responsible for health.' This drove his decision to become a brain surgeon. Then, during his hospital training, he was looking after a patient with serious burns who suddenly suffered severe inflammation. 'She was an ii-month-old baby girl called Janice who died in my arms.'

These traumatic moments made him a neurosurgeon who thinks a lot about inflammation. He believes it was this perspective that enabled him to interpret the results of an accidental experiment in a new way.

In the late 1990s, Tracey was experimenting with a rat's brain. 'We'd injected an anti-inflammatory drug into the brain because we were studying the beneficial effect of blocking inflammation during a stroke,' he recalls. 'We were surprised to find that when the drug was present in the brain, it also blocked inflammation in the spleen and in other organs in the rest of the body. Yet the amount of drug we'd injected was far too small to have got into the bloodstream and travelled to the rest of the body.'

After months puzzling over this, he finally hit upon the idea that the brain might be using the nervous system – specifically the vagus nerve – to tell the spleen to switch off inflammation everywhere.

It was an extraordinary idea – if Tracey was right, inflammation in body tissues was being directly regulated by the brain. Communication between the immune system's specialist cells in our organs and bloodstream and the electrical connections of the nervous system had been considered impossible. Now Tracey was apparently discovering that the two systems were intricately linked.

The first critical test of this exciting hypothesis was to cut the vagus nerve. When Tracey and his team did, injecting the anti-inflammatory drug into the brain no longer had an effect on the rest of the body. The second test was to stimulate the nerve without any drug in the system. 'Because the vagus nerve, like all nerves, communicates information through electrical signals, it meant that we should be able to replicate the experiment by putting a nerve stimulator on the vagus nerve in the brainstem to block inflammation in the spleen,' he explains. 'That's what we did and that was the breakthrough experiment.'

§ § §

Operating far below the level of our conscious minds, the vagus nerve is vital for keeping our bodies healthy. It is an essential part of the parasympathetic nervous system, which is responsible for calming organs after the stressed 'fight-or-flight' adrenaline response to danger. Not all vagus nerves are the same, however: some people have stronger vagus activity, which means their bodies can relax faster after a stress.

The strength of your vagus response is known as your vagal tone and it can be determined by using an electrocardiogram to measure heart rate. Every time you breathe in, your heart beats faster in order to speed the flow of oxygenated blood around your body. Breathe out and your heart rate slows. This variability is one of many things regulated by the vagus nerve, which is active when you breathe out but suppressed when you breathe in, so the bigger your difference in heart rate when breathing in and out, the higher your vagal tone.

Research shows that a high vagal tone makes your body better at regulating blood glucose levels, reducing the likelihood of diabetes, stroke and cardiovascular disease. Low vagal tone, however, has been associated with chronic inflammation. As part of the immune system, inflammation has a useful role helping the body to heal after an injury, for example, but it can damage organs and blood vessels if it persists when it is not needed. One of the vagus nerve's jobs is to reset the immune system and switch off production of proteins that fuel inflammation. Low vagal tone means this regulation is less effective and inflammation can become excessive, such as in Maria Vrind's rheumatoid arthritis or in toxic shock syndrome, which Kevin Tracey believes killed little Janice.

Having found evidence of a role for the vagus in a range of chronic inflammatory diseases, including rheumatoid arthritis, Tracey and his colleagues wanted to see if it could become a possible route for treatment. The vagus nerve works as a two-way messenger, passing electrochemical signals between the organs and the brain. In chronic inflammatory disease, Tracey figured, messages from the brain telling the spleen to switch off production of a particular inflammatory protein, tumour necrosis factor (TNF), weren't being sent. Perhaps the signals could be boosted?

He spent the next decade meticulously mapping all the neural pathways involved in regulating TNF, from the brainstem to the mitochondria inside all our cells. Eventually, with a robust understanding of how the vagus nerve controlled inflammation, Tracey was ready to test whether it was possible to intervene in human disease.

§§§

In the summer of 2011, Maria Vrind saw a newspaper advertisement calling for people with severe rheumatoid arthritis to volunteer for a clinical trial. Taking part would involve being fitted with an electrical implant directly connected to the vagus nerve. 'I called them immediately,' she says. 'I didn't want to be on anticancer drugs my whole life; it's bad for your organs and not good long-term.'

Tracey had designed the trial with his collaborator, Paul-Peter Tak, professor of rheumatology at the University of Amsterdam. Tak had long been searching for an alternative to strong drugs that suppress the immune system to treat rheumatoid arthritis. 'The body's immune response only becomes a problem when it attacks your own body rather than alien cells, or when it is chronic,' he reasoned. 'So the question becomes: how can we enhance the body's switch-off mechanism? How can we drive resolution?'

When Tracey called him to suggest stimulating the vagus nerve might be the answer by switching off production of TNF, Tak quickly saw the potential and was enthusiastic to see if it would work. Vagal nerve stimulation had already been approved in humans for epilepsy, so getting approval for an arthritis trial would be relatively straightforward. A more serious potential hurdle was whether people used to taking drugs for their condition would be willing to undergo an operation to implant a device inside their body: 'There was a big question mark about whether patients would accept a neuroelectric device like a pacemaker,' Tak says.

He needn't have worried. More than a thousand people expressed interest in the procedure, far more than were needed for the trial. In November 2011, Vrind was the first of 20 Dutch patients to be operated on.

'They put the pacemaker on the left-hand side of my chest, with wires that go up and attach to the vagus nerve in my throat,' she says. 'I waited two weeks while the area healed, and then the doctors switched it on and adjusted the settings for me.'

She was given a magnet to swipe across her throat six times a day, activating the implant and stimulating her vagus nerve for 30 seconds at a time. The hope was that this would reduce the inflammatory response in her spleen. As Vrind and the other trial participants were sent home, it became a waiting game for Tracey, Tak and the team to see if the theory, lab studies and animal trials would bear fruit in real patients. 'We hoped that for some, there would be an easing of their symptoms – perhaps their joints would become a little less painful,' Tak says.

At first, Vrind was a bit too eager for a miracle cure. She immediately stopped taking her pills, but her symptoms came back so badly that she was bedridden and in terrible pain. She went back on the drugs and they were gradually reduced over a week instead.

And then the extraordinary happened: Vrind experienced a recovery more remarkable than she or the scientists had dared hope for. 'Within a few weeks, I was in a great condition,' she says. 'I could walk again and cycle, I started ice-skating again and got back to my gymnastics. I feel so much better.' She is still taking methotrexate, which she will need at a low dose for the rest of her life, but at 68, semi-retired Vrind now plays and teaches seniors' volleyball a couple of hours a week, cycles for at least an hour every day, does gymnastics, and plays with her eight grandchildren.

Other patients on the trial had similar transformative experiences. The results are still being prepared for publication but Tak says more than half of the patients showed significant improvement and around one-third are in remission – in effect cured of their rheumatoid arthritis. Sixteen of the 20 patients on the trial not only felt better, but measures of inflammation in their blood also went down. Some are now entirely drug-free. Even those who have not experienced clinically significant improvements with the implant insist it helps them; nobody wants it removed.

'We have shown very clear trends with stimulation of three minutes a day,' Tak says. 'When we discontinued stimulation, you could see disease came back again and levels of TNF in the blood went up. We restarted stimulation, and it normalised again.'

Tak suspects that patients will continue to need vagal nerve stimulation for life. But unlike the drugs, which work by preventing production of immune cells and proteins such as TNF, vagal nerve stimulation seems to restore the body's natural balance. It reduces the over-production of TNF that causes chronic inflammation but does not affect healthy immune function, so the body can respond normally to infection.

'I'm really glad I got into the trial,' says Vrind. 'It's been more than three years now since the implant and my symptoms haven't returned. At first I felt a pain in my head and throat when I used it, but within a couple of days, it stopped. Now I don't feel anything except a tightness in my throat and my voice trembles while it's working.

'I have occasional stiffness or a little pain in my knee sometimes but it's gone in a couple of hours. I don't have any side-effects from the implant, like I had with the drugs, and the effect is not wearing off, like it did with the drugs.'

§ § §

But what about people who just have low vagal tone, whose physical and mental health could benefit from giving it a boost? Low vagal tone is associated with a range of health risks, whereas people with high vagal tone are not just healthier, they're also socially and psychologically stronger – better able to concentrate and remember things, happier and less likely to be depressed, more empathetic and more likely to have close friendships.

Twin studies show that to a certain extent, vagal tone is genetically predetermined – some people are born luckier than others. But low vagal tone is more prevalent in those with certain lifestyles – people who do little exercise, for example. This led psychologists at the University of North Carolina at Chapel Hill to wonder if the relationship between vagal tone and wellbeing could be harnessed without the need for implants.

In 2010, Barbara Fredrickson and Bethany Kok recruited around 70 university staff members for an experiment. Each volunteer was asked to record the strength of emotions they felt every day. Vagal tone was measured at the beginning of the experiment and at the end, nine weeks later. As part of the experiment, half of the participants were taught a meditation technique to promote feelings of goodwill towards themselves and others.

Those who meditated showed a significant rise in vagal tone, which was associated with reported increases in positive emotions. 'That was the first experimental evidence that if you increased positive emotions and that led to increased social closeness, then vagal tone changed,' Kok says.

Now at the Max Planck Institute in Germany, Kok is conducting a much larger trial to see if the results they found can be replicated. If so, vagal tone could one day be used as a diagnostic tool. In a way, it already is. 'Hospitals already track heart-rate variability – vagal tone – in patients that have had a heart attack,' she says, 'because it is known that having low variability is a risk factor.'

The implications of being able to simply and cheaply improve vagal tone, and so relieve major public health burdens such as cardiovascular conditions and diabetes, are enormous. It has the potential to completely change how we view disease. If visiting your GP involved a check on your vagal tone as easily as we test blood pressure, for example, you could be prescribed therapies to improve it. But this is still a long way off: 'We don't even know yet what a healthy vagal tone looks like,' cautions Kok. 'We're just looking at ranges, we don't have precise measurements like we do for blood pressure.'

However the technology develops, our understanding of how the body manages disease has changed for ever. 'It's become increasingly clear that we can't see organ systems in isolation, like we did in the past,' says Paul-Peter Tak. 'We just looked at the immune system and therefore we have medicines that target the immune system.

'But it's very clear that the human is one entity: mind and body are one. It sounds logical but it's not how we looked at it before. We didn't have the science to agree with what may seem intuitive. Now we have new data and new insights.'

And Maria Vrind, who despite severe rheumatoid arthritis can now cycle pain-free around Volendam, has a new lease of life: 'It's not a miracle – they told me how it works through electrical impulses – but it feels magical. I don't want them to remove it ever. I have my life back!'

*This story was first published on 26 May 2015
by Wellcome on mosaicscience.com*

Could virtual reality headsets relieve pain?

■ Jo Marchant

'It's like a crawly feeling inside,' says Judy*. 'You get hot, then chilled, and you feel like you want to run away.' The 57-year-old has short dark-grey hair and a haunted expression. She's breathless and sits with her right leg balanced up on her walking stick, rocking it back and forth as she speaks.

Judy explains that she suffers from constant, debilitating pain: arthritis, back problems, fibromyalgia and daily migraines. She was a manager at a major electronics company until 2008, but can no longer work. She often hurts too much even to make it out of bed.

She's taking around 20 different medications each day, including painkillers, antidepressants, sedatives and a skin patch containing a high dose of the opioid drug fentanyl, which she says did not significantly help her pain and which she's now trying to come off. Her physician has been tapering the dose for months, so in addition to her pain she suffers withdrawal symptoms: the chills and crawling dread. Then her clinic announced that it would no longer prescribe any opioids at all, the unintended result of new, stricter measures aimed at clamping down on opioid abuse. Faced with losing access to the drug on which she is physically dependent, she has come to another clinic, Pain Consultants of East Tennessee (PCET) in Knoxville, desperate for help.

Ted Jones, the attending clinician, calls patients like Judy 'refugees'. He says that he sees 'tons' of similar cases. Over 100 million Americans suffer long-term pain. Now they find themselves at the epicentre of two colliding health catastrophes in the USA: chronic pain and opioid abuse.

Over the last few decades, US doctors have tackled constant pain problems by prescribing ever-higher levels of opioid pain-killers – drugs such as hydrocodone and oxycodone, which belong to the same chemical family as morphine and heroin. These medications have turned out to be less effective for treating chronic pain than thought – and far more addictive. The surge in prescriptions has fed spiralling levels of opioid abuse and tens of thousands of overdose deaths.

Efforts to curb opioid prescriptions and abuse are starting to work. But with the spectacular failure of a drug-centric approach to treating chronic pain, doctors desperately need alternative ways to fight a condition that blights millions of lives. Jones is trying one, seemingly unlikely technological solution: virtual reality.

§ § §

Since the 1990s, opioid prescriptions have tripled in the USA, which has less than 5 per cent of the world's population but now uses more than 80 per cent of the world's opioid supply – with devastating consequences.

Opioid painkillers were previously used only in specialist cases, such as short-term pain after surgery or terminal cancer pain. But a campaign by the American Pain Society arguing for pain to be treated more aggressively – as well as marketing from pharmaceutical companies, claiming that newly approved opioid drugs such as OxyContin were effective and non-addictive – resulted in doctors prescribing them much more widely for many different types of chronic pain. 'We were told pain was undertreated,' says Joe Browder, a physician and senior partner of PCET. 'There was no upper limit.'

The new drugs turned out to be highly addictive. In 2015, more than 16,000 people died from overdoses of prescription painkillers, in what figures from the US Centers for Disease Control and Prevention (CDC) show is the USA's worst ever drug epidemic.

Tennessee is one of the worst-hit states in terms of opiate prescriptions and overdose deaths. Part of the state lies in the Appalachian mountains – a rural region that has long struggled with economic deprivation and lack of jobs, caught in what Karen Pershing of the Metro Drug Coalition in Knoxville describes as a 'culture of addiction and poverty'. The narcotic pills prescribed by doctors have flowed easily from pain clinics into the surrounding community. 'People couldn't afford street drugs,' says Pershing. Suddenly they had access to prescription medications that provided the same psychological escape.

In particular, there was an explosion of unregulated but legal pain clinics known as 'pill mills'. All that was needed was for the director of a clinic to be a qualified medical doctor. Staff could then hand out prescriptions for opioid painkillers in return for cash. 'Every time you turned around there was another one,' says Pershing. Some of them were in old houses, with handwritten signs. People would be in and out in a few minutes, selling on the phone as soon as they got out.

§ § §

Over the last five years or so, there have been a series of measures aimed at curbing opioid prescriptions and abuse, from requiring pain clinics to be run by doctors certified in pain treatment, to encouraging the public to report suspicious activity by text message. In some states, including Tennessee, physicians must regularly count patients' pills to ensure they're not taking too many or selling them on, and face criminal charges if patients abuse their drugs. Not surprisingly, some clinics now refuse to prescribe opioids at all.

PCET is in one of the hardest-hit areas. According to the US Senate Committee on Health, Education, Labor, and Pensions, Tennessee ranks third in the nation for prescription drug abuse, with 5 per cent of the population abusing these drugs. Here in Knox County, there were 133 deaths from opioid overdoses in 2014, a higher death rate than any other county in the state.

Physicians at PCET do still use opioid drugs, though with strict precautions. For example, Jones runs courses to help patients treat their pills responsibly and stay within the law. 'Take $1,000 cash and put it with your pain meds,' he tells them. 'Or put a loaded gun there. If cash or a gun isn't safe there, your pain meds aren't safe there.' (Rule of thumb: 'If they leave you some, it was family. If they clean you out, it wasn't.')

He also interviews new patients like Judy to assess how likely they are to abuse their medication. Other refugees chasing opioids this morning include Scott*, a balding teetotaller in his 60s who runs a barbecue joint. Injuries from working at rodeos and horse shows in his youth as well as nerve damage from recent chemotherapy mean 'if I'm moving, I'm hurting'. Meanwhile Debra*, a 65-year-old widow with turquoise shorts, suffers from lupus, arthritis, fibromyalgia, diabetic neuropathy and sciatica, not to mention anxiety and depression. 'I hurt everywhere,' she says. She hides her drugs around the house but says her drug-addict daughter still finds them. Jones puts her at medium risk for misusing. She'll still get opioids, but will have a drug screen every month or two to make sure she's not taking too much.

Measures like these are starting to work. Between 2012 and 2015, there was a 12–18 per cent drop in the amount of opiate painkillers prescribed nationally. But that leaves many patients struggling to cope without these drugs, while more Americans than ever – some estimates say over 100 million – are in chronic pain, costing the US economy around $600 billion every year. (Although other countries haven't seen such a large spike in opioid use as the USA, chronic pain is a huge and growing problem globally, affecting 15–20 per cent of adults in industrialised nations.)

Despite taking their drugs responsibly, the patients Jones sees today are all still desperate with pain. Listening to their stories, the overwhelming impression is of chronic pain not as a discrete physical condition that can be solved with a pill, but as a diffuse suffering that grows and spreads around the body, evading the efforts of ever-increasing medications and often entwined with arduous life histories and psychiatric issues such as anxiety and depression. 'Pain and anxiety are intertwined,' confirms Jones. 'It's hard to know when you are treating anxiety or pain.' Physicians often ask patients to rate their pain out of ten, but 'the pain score is really a distress score,' he reckons. 'It's always a seven. It's a measure of depression, unhappiness and pessimism.'

Browder says he realised more than a decade ago that ever-higher drug doses weren't the answer. Patients may not have been reporting as much pain, but the drugs weren't doing anything, he says. Instead of trying to eliminate pain with narcotics and sedatives, Browder and his colleagues decided to prioritise patient function. 'It's all about getting people to do more in their life with the pain they have,' explains PCET medical director James Choo. 'How do you get people to do more, to live more, to be more like the people they were before they started hurting?'

They started to reduce the amount of opioids they prescribed, emphasising other medical interventions such as steroid injections, nerve ablation and spinal cord stimulators. The clinic also started offering physical therapy and occupational therapy, while an embedded team of psychologists, including Jones, offers counselling as well as advice in proven pain-relief techniques such as cognitive behavioural therapy, distraction and relaxation mindfulness.

Persuading patients to embrace this more diverse approach isn't easy, however. Relaxation techniques work well, but they take practice, and Jones says they've struggled to attract patients to multi-session courses on techniques for coping with pain. A few people will do it, but in general life gets in the way, he says, adding that there are a million reasons why they don't come. People don't want programmes, agrees Choo. 'They just want to take a pill.'

§ § §

One day last winter, Jones came to work to find an empty clinic; heavy snowfall had kept many patients at home. He filled time by surfing the internet, and stumbled across the website of a technology start-up called Firsthand Technology, based 2,000 miles from Knoxville in San Francisco.

For the company's CEO, Howard Rose, virtual reality – or VR – is nothing less than a superpower, or as the company website puts it, a 'high-bandwidth channel' into our brains that can transform how we see ourselves and the world. Within a couple of generations, he predicts, VR will be woven into every aspect of our lives. He's starting with how we manage pain.

Rose began working in VR more than 20 years ago, at the Human Interface Technology Lab (HITLab) at the University of Washington in Seattle. The lab was founded by Tom Furness, who had pioneered the use of VR when working for the US Air Force in the 1960s. Under Furness, Rose and his colleagues developed civilian uses of the technology, creating VR worlds for everything from treating spider phobias to teaching Japanese. One of the most successful products to come out of the lab was SnowWorld, developed by cognitive psychologist Hunter Hoffman to ease the pain of patients with severe burns.

Burns patients have to undergo regular wound-care sessions so painful that they can be excruciating even with high doses of painkillers. SnowWorld was designed as a kind of souped-up distraction method for use during these sessions, to divert patients' attention away from their pain. Adapted from flight simulation software, it creates the experience of flying through a virtual ice canyon while exchanging snowballs with penguins and snowmen.

Over the past ten years or so Hoffman and his colleagues have shown in several trials, including on army veterans burned by explosive devices in Iraq and Afghanistan, that this works. Playing SnowWorld during wound-care sessions eases patients' reported pain up to 50 per cent in addition to the pain relief they get from drugs – significantly better than other forms of distraction, such as music or video games. Studies also show that SnowWorld reduces activity in areas of the brain associated with pain perception.

The researchers believe that the sense of immersion created by VR – feeling physically present in the virtual location – is crucial. 'It works because it tricks your senses into perceiving that the computer-generated environment is real,' says Rose. 'VR becomes a place you are, not something that you are watching.'

Thanks to Hoffman's research, SnowWorld has become famous, featuring in numerous press reports and even in an off-Broadway play starring Hollywood actress Mamie Gummer that is now showing at the National Theatre in London. Research interest has grown too, and VR immersion has since been shown to reduce reported pain and distress during a range of medical and dental procedures, from chemotherapy to taking blood.

Meanwhile, Rose and his colleague Ari Hollander left academia to form Firsthand Technology. SnowWorld is a research tool; they wanted to develop commercial products. They built Cool!, which Rose describes as 'a sort of next generation of what we learned in SnowWorld'. Featuring more interactivity and a wider variety of environments, it's more open-ended, he says, 'a kind of playground'.

I tried it myself at the PCET clinic. A few seconds after slipping on the headset, I was floating along a river with grassy banks. There were mountains in the distance, and a blue sky with scattered clouds. Along the edge of the water, fluffy brown otters stood greeting me on their hind legs. Using two hand controllers, I threw fish to them and they rolled over in delight, changing colour to zebra stripes or flamingo pink. Then I launched huge bubbles that bounced pleasingly around the landscape.

The graphics were detailed, if a little cartoonish, but what struck me most was the physicality of the experience – to my brain, this otter-inhabited world wasn't simply something I was watching, but a place that I was actually in. When I passed under a rocky bridge, I flinched. When snow fell, I felt the exhilaration of clear, cold air.

VR is undoubtedly effective at shifting attention, but Rose argues that it works on other levels too. 'We know that if people feel anxious and helpless then their suffering from the pain is much greater,' he says. Mentally taking people to a distant, safe place reduces their anxiety, he says, while interactivity – the ability to move around an environment and throw snowballs, for example – helps them to feel more in control.

He'd wondered whether these attributes might help patients with chronic conditions too – those suffering from pain, anxiety and helplessness in daily life. Then he received an email from Jones in Tennessee. 'We've got patients, you've got a product,' said Jones. A few months later, Rose flew to Knoxville and dropped off his equipment. But would it work?

§ § §

Christine* smiles. 'Mmmm,' she says. 'You can almost feel the petals hitting you.' The 69-year-old is sitting in Jones's green leather armchair, her eyes hidden by a bulky black visor.

Jones's laptop shows Christine's field of view as she floats down the stream. She looks around from the otters to some mysterious balls of dancing coloured flame, then scrunches up her face as she brushes past the blossom-laden branches of a grove of cherry trees.

Christine has led an active life: she used to run a tour company in Mexico, before becoming a chef at her local synagogue. But despite her 'very full' life – which has included regular childcare – she now finds it difficult to focus on anything but her pain. Since 2007 she has suffered from an autoimmune condition that attacks her nerves, causing burning pain down both legs and across the soles of her feet. She is also recovering from back surgery. She has resisted opioids and stopped taking two other drugs, Lyrica and Cymbalta, because they interrupted her thinking and speech. 'It spoiled my life,' she says. Now she gets by on Advil (ibuprofen) and lidocaine patches, which she wears on her legs during the day and on her feet at night, but the pain never goes away. 'It's all you think about,' she says.

When she arrives in Jones's office, Christine rates her pain as 7/10. Then she puts on the headset and headphones for her third session of VR. 'I'm already relaxing,' she says. After the cherry trees she follows the river into a cave, its rocky walls covered with strange drawings and sparkling gems. How's your pain, I ask. 'Oh,' she says, as if surprised to be reminded of it. 'Zero. It's gone.'

Compared to other forms of distraction, such as colouring or watching TV, VR 'works a whole lot better,' Jones says. 'It grabs your attention. You just put the helmet on and you are gone.'

Jones has recently completed two small clinical trials of Cool!, which together involved 40 participants receiving between them around 60 sessions of VR. Only one person didn't report reduced pain, he says. Overall, the patients reported that their pain fell by 60–75 per cent (compared to baseline) during their

VR session, and by 30–50 per cent immediately afterwards. The best morphine does is 30 per cent.

It's early research, but a few other studies of chronic pain have found similar results. Earlier this year, Diane Gromala of the Pain Studies Lab at Simon Fraser University, Canada, and her colleagues reported on a VR game called Cryoslide – inspired by SnowWorld – which involves sliding through a snowy landscape and icy cave while throwing snowballs at fantasy creatures. When chronic pain patients played it for ten minutes, their pain reduced significantly compared to those asked to use other distraction strategies such as meditating, reading or playing games on a phone. Elsewhere, Brenda Wiederhold and colleagues at the Virtual Reality Medical Center in San Diego found that chronic pain patients immersed in virtual scenes such as forests, beaches and mountains reported significantly reduced pain compared to baseline. They also had a reduced heart rate and raised skin temperature, suggesting that they were more relaxed. According to the published paper, patients said things like: 'This is the first pain relief I have had in three years' and 'I was so busy playing the game, I forgot about my pain'.

No peer-reviewed studies have yet investigated whether VR is helpful for managing chronic pain long-term. But the research so far provides the first proof of principle that VR can ease pain in chronic conditions. It isn't possible or desirable to immerse patients all the time, of course, but even short sessions might provide welcome respite for patients who are otherwise always in pain. Jones says that compared to techniques such as mindfulness, which require long-term effort and training, 'with VR you put a visor on their head and they're there'.

He hopes that the effects last at least a little longer than the session. In his trials, patients reported that their pain relief lasted for between 2 and 48 hours afterwards. This supports the idea that it might be triggering the release of pain-relieving endorphins, for example, which ease symptoms even after the headset is removed. Perhaps it also demonstrates to patients that their pain 'is not intractable, it can be influenced,' suggests Browder. In other words, it shows people with grinding, constant pain that there is hope.

That interpretation chimes with the accounts of Jones's patients. Cindy* is 54 years old and suffers from non-Hodgkin's lymphoma as well as familial pancreatitis, a condition that killed her brother and sister and which causes severe stomach pains and nerve damage. She's on 17 medications three times a day. After the VR session, she's visibly calmer and a noticeable hand tremor has temporarily eased. The VR 'is an escape from pain reality,' she says. Her husband agrees: 'After she does the VR, her mood lightens,' he says. 'She's more upbeat towards the world.'

Andrew*, a tattoo-covered ex-Green Beret, suffers pain as a result of nerve damage from exposure to chemical weapons in Iraq, combined with back problems 'from jumping out of planes and carrying a heavy rucksack'. Opioid painkillers haven't helped and he has often contemplated suicide, but the VR takes it all away, he says. 'It made me feel the way a man feels after achieving an orgasm.'

§ § §

There is increasing interest in using VR to teach long-term coping skills that patients can use in their daily lives, or as Rose puts it, 'to change that relationship with the pain'. One way to do this is to combine the VR with biofeedback, creating environments that change according to a person's physiological state. For example, Gromala and her colleagues have developed a virtual environment aimed at teaching mindfulness-based stress reduction (MBSR). There is robust evidence that MBSR reduces chronic pain, but it takes time and effort to learn. Gromala's 'virtual meditative walk' takes users along forest trails as they listen to a guided meditation. Biosensors track their arousal level, and modify the weather in response: light fog recedes as the patient relaxes, or thickens if they become more stressed.

In one trial, patients who tried the virtual walk had significantly reduced pain compared to those who listened only to the audio track. Gromala and her colleagues now need to test whether regular sessions enable users to better control stress and pain in everyday life. They don't just want to distract users from their pain, they write in a paper, but to 'arm them with a learned capacity to modulate it'.

Meanwhile, Jones is discussing with Rose the idea of a 'pain tamer', in which by reducing their stress levels, patients can make a virtual monster – representing their pain – shrink and disappear. 'They learn to relax themselves,' says Jones. 'It's not so much distraction as learning how to calm yourself in the face of an angry adversary.' Rose and Gromala envisage such virtual worlds running on a variety of platforms – where patients might experience high-quality immersive VR in a clinic, supplemented at home by daily top-up sessions on tablets or phones.

'People talk about waves of VR,' says Rose. 'They were always little ripples. This is a tsunami.' When SnowWorld was first developed in 1990s, Hoffman and his colleagues used a supercomputer and heavy helmet that together cost $90,000. Today, users can buy lightweight high-resolution headsets such as the Oculus Rift, which is the size of a scuba mask, for around $600. HTC and Sony offer similarly priced devices, while vastly cheaper sets by Google and Samsung don't have such impressive displays but can run off tablets and smartphones slotted into cardboard or plastic headgear. For Rose, all this amounts to nothing less than a revolution in medicine.

'The treatment model of medicine is you come in and stuff is done to you,' he says. VR, on the other hand, shifts responsibility and control to the patient. By exploring virtual worlds, those crippled by anxiety or pain may gain the skills, strength and courage they need to transform their experience of the real one.

There are several obstacles to its use becoming routine. One is a reluctance among many hospitals and clinics to adopt unfamiliar technology. Even where the research evidence is strong – in burns patients, for example – VR pain relief is still not widely used. Jones describes how PCET raised funds for a nearby children's hospital, and demonstrated to one of the nurses how VR could help children going through medical procedures. She wasn't impressed. 'It's frustrating,' says Jones. 'We gave her $15,000 and she's going to buy more colouring books.'

Developing VR into a routine treatment is also going to require new models of funding. So far, Jones has paid for his clinical trials himself. 'It's my hobby,' he says. 'I don't have a boat.' Beyond those, he actively discourages his physician colleagues from referring patients for VR, despite the fact that he is convinced it would help, because he can't bill for it. VR pain relief needs a 'champion' he says, who will fund trials and convince insurance companies to take it seriously. 'The VR industry has got some work to do.'

Meanwhile, he'd just taken delivery of Firsthand Technology's latest virtual world, called Glow. I tried it, and found myself standing in a moonlit forest clearing in front of what looked like a carved stone altar. Fireflies danced through the air. At Jones's instruction I held out my hands and focused on breathing calmly, slowly. A sensor on my fingertip relayed my heart rate to the computer so that as I relaxed, hundreds of fireflies settled between my hands, eventually creating a glowing sphere.

Instead of an experience that you play with, says Rose, 'this is having the world react to you.' One application is customising an experience to someone's changing pain. For example, for women in labour, Rose suggests that a virtual world might offer 'superpowers and catharsis and blowing stuff up' during contractions, while promoting relaxation in between.

For Jones, there won't ever be one answer to chronic pain. But VR could be another tool. As well as providing a non-pharmacological alternative to treating pain, it might help people who are on opioids to minimise their dose, giving them something active they can do instead of popping extra pills. 'With opioids, you just take it, sit and wait.' That's why people end up taking more pills than they need – they just want to do something to help their pain.'

This story was first published on 25 April 2017
by Wellcome on mosaicscience.com

Patients' real names have been anonymised in this article.

What it means to be homesick in the 21st Century

■ John Osborne

I'm standing outside a cottage in the Vienna woods. It's where I used to live and this is the first time I've been back since I left ten years ago. The roads are steep here, and as I walked up the hill from the tram stop the slope felt reassuringly familiar. I recognised the ache in the back of my legs that told me I was nearly there.

Saturnweg, Merkurweg, Jupiterweg. The streets take their names from the solar system, and that seems appropriate for a place that felt so alien and far away when I first arrived. There is a romance to the planets, though, and I feel the same about the alignment of these streets. In the distance is the most spectacular view of the hills. And here, the little cottage with the green gate. The only house on the street without a swimming pool.

I'm not surprised I was homesick when I first arrived here. The 22-year-old version of me must have been completely out of his depth. I didn't keep a diary back then, but if I did, for day one I'm sure I'd have just written 'Oh dear'.

§§§

The first time I ever remember feeling homesick was at Cub [Scout] camp. Being away from home for the first time is a terrifying experience, but sometimes there are grown-ups in woggles to look after you. In that Austrian cottage, inside the front room, I felt the most overwhelming homesickness I have ever experienced: it was a physical pain, and it lasted for a couple of weeks. That seems like no time at all now, looking back, but at the time it felt like it would last longer than the Austro-Hungarian Empire.Luckily, as is predominantly the case, the homesickness didn't last. But sometimes homesickness can linger, exacerbated by society's refusal to address it directly as an issue. For centuries, the way to deal with homesickness has been to pretend it does not exist. Susan Matt, author of the book Homesickness: An American history, writes: 'Because homesickness is absent from modern accounts of the past, it is seen as an illegitimate emotion in the present.'

Portrayals of early American settlers suggest that they didn't have a problem with homesickness. It was in almost direct opposition to the very essence of what the new country stood for – freedom of movement.

Matt points out, however, that a yearning for what the pioneers had left behind was clearly present: 'The paths of homesick migrants can be traced through the repetition of place-names across the American landscape,' she writes. 'English town names were transplanted to New England; subsequent generations settling in the Midwest and West carried these names with them and tried to re-establish a sense of place by affixing old names to new locales.'

What use, if any, is homesickness? 'Its purpose is the same today as it has been for millions of years – to deter us from leaving supportive groups and environments,' writes Mark Leary,

Professor of Psychology and Neuroscience at Duke University in the USA, in Duke Magazine. 'Homesickness would have been relatively uncommon, occurring only when individuals were separated from supportive, familiar people.'

His is the best definition of homesickness I have found: a feeling of wanting to be back with our tribe. They are hard to come by, these supportive, familiar people. It's no surprise we feel lost when we move away from them.

It was my parents that I was desperate to contact when I first arrived in Vienna. They were my supportive, familiar people. My tribe. It wasn't that I missed them; I just wanted them to know that I had arrived and started to settle in. I thought maybe they'd be worried, and I wanted to say that everything was going to be okay.

My landlady sorted me out with a key, told me what day the binmen came and showed me the cord I needed to pull to make the shower work. From the moment she closed the door behind her, I was completely alone. The only thing I wanted to do was find a phone box and talk to my parents. I could call, say hello and then get on with starting my new Viennese life.

It was getting late. I'd been walking for so long but, after almost giving up, I finally found a phone box. I dialled my parents' number, proud of myself for remembering the international dialling code, but the line was dead and my coins were swallowed. All my Euros were gone. That was the only pay phone. Now there was no way to contact them. I felt devastated.

I still think about that first night and the early days living completely alone in a country where I knew no one; I feel bad for all the people across the world who have gone through something similar. I just wanted to hear a familiar voice. I just wanted to tell someone, 'Don't worry, I'm okay'. It took me a long time to recover from that less-than-promising start: I felt sadder than I had ever felt before. My plan of a new life in Vienna had failed. I didn't even unpack my suitcase – there didn't seem to be any point. There was no way I would be staying.

§ § §

I wanted to know why I felt this way. Professor Doreen Massey, a recently retired social geographer who has specialised in globalisation and the 'conceptualisation of place', explained to me that one problem for Western men is our idealisation of home. 'A lot of writers, in lots of genres, you will find an idealisation of home,' she says. 'That kind of idealisation and romanticisation… is at the same time denigration, ie 'It was lovely, it was unproblematical, it was contained'…which of course it never was, and it's not as it was any more.'

Homesickness often feels like unrequited love because we have such a connection with places we are fond of. We build this perfect image of the person or place we are missing. We remember the best-case scenario as the everyday occurrence. Our brains filter out the bad bits, focusing on the day everything was perfect. To think of not being there, or not being with the person, any more makes us feel so helpless.

But maybe homesickness, like lovesickness, can be a good thing – perhaps there is a positive way of looking at it. Susanna Barry, a Senior Program Manager at MIT Medical (which provides healthcare for the Massachusetts Institute of Technology), specialises in stress management and sleep health. Speaking on a Conversations with MIT Medical podcast about homesickness, she gives advice on how to tolerate new experiences and suggests that thriving on them is the silver lining of homesickness.

'Post-homesickness growth is very real and very empowering,' she says. 'It's always true to say 'This too shall pass'.'

But what if it doesn't pass? Much of what is written on the subject of homesickness relates to the luxury of being away from home. It's mainly articles of advice for university students that include snippets such as 'Homesickness is a normal part of college students' development toward adulthood' and 'When you enjoy your studies, you'll probably feel less homesick'. But for many people moving away from home, there is no choice.

Massey and I discuss migrants who are forced to start from scratch somewhere new, who are often alone and frequently have no connection to their new location. To talk to someone about homesickness, she says, you need to know the nature of their journey. What is the transition that has been made?

'If I have home it's the North-west,' she tells me from her London base. 'Manchester, Liverpool. That's where I do feel home. I don't get homesickness for it, but I do think of it as my patch... If I was talking to somebody who was homesick for Liverpool, say, I would talk...in a very different way than if I was talking to a refugee who could not go home, whose place had been devastated.'

If I or Professor Massey feel like we need to visit somewhere we remember from childhood, we can get on a train and take a journey there. It's not the same if the place you called home has been erased.

§ § §

The British Red Cross was set up in 1870 and today gives help and advice to people in crisis. They assess whether refugees and asylum seekers are eligible for support and accommodation.

Jane, one of 27,000 Red Cross volunteers across the UK, tells me over the phone that she's just seen a client who was definitely homesick. 'Last time I saw him he was in tears because he thought he might never see his mother again.' In fact, she says, he probably won't. He had been put in prison in his home country and when he was released, he felt he couldn't go back, so came to the UK.

The volunteers meet many people who are badly in need of help. Jane and I discuss Maslow's hierarchy of needs, a theory that's used in psychology and business and concerns the growth of individuals. 'We are working with clients who are working on the very basic [physiological] needs: they have fresh air to breathe, but that's about all they've got.

'I wonder where homesickness fits in with that. Either our clients are incredibly resilient or when you are struggling – thinking 'Where am I going to sleep tonight?' or 'Where am I going to find something to eat tonight?' – the homesickness wouldn't overwhelm you so much.'

Adeela Bainbridge is the Red Cross's International Family Tracing Co-ordinator for Cambridgeshire, Norfolk and Suffolk. I asked her what emotional condition people are in when they arrive in the UK. '[People who] come from Gaza, Syria or Rwanda often don't want to acknowledge what they have left behind, and I can understand that,' she says. 'It's too painful to think about what you're missing or what those people are going through.'

She talks me through her role of helping any refugee who comes to see her. If necessary, she conducts a tracing enquiry; the Red Cross accepts tracing enquiries from people living in the UK who have been separated from relatives as a result of conflict, disaster or migration.

She tells me that she asks for very precise details ('Tell me what your house looked like. Are there any particular trees in the area?') and never knows what might emerge. 'We actually found a person because they had a mulberry tree growing in their garden,' she tells me. 'The tiniest details evoke such memories that people often break down and cry. They have such a keen sense of loss.'

As tracing coordinator Bainbridge works with people at a variety of ages, from their early teens to their late eighties. They have often been displaced because of natural disasters, or past or present conflict. 'What does emerge,' Bainbridge says, 'are the milestones they share with nobody but [that] are significant. If they have achieved a qualification or are celebrating another birthday... It brings it home to them.

Sometimes it's a physical thing, about food or music. 'The young people I work with say 'I just miss my mother' or 'I'm afraid of the dark'. It's heartbreaking.'

The suddenness of displacement means that people leave everything behind – their family, their way of life. 'When they come here, either because of the language barrier or the way asylum works, that loss... is so much more enhanced.'

§ § §

Part of adjusting to a new place is discovering ways to use your skills. Bainbridge told me that once refugees have settled after their initial relocation, they might begin to write poetry again, or make music, or find a place to go running and realise that they now have an exercise routine. Things start to feel familiar.

In Vienna, I eventually unpacked my suitcase. I even found a favourite place, a bar called 1516. There were three of us who used to go there together: me; Wolfie, who taught physics at the same school that I worked at; and Wolfie's mate Liam, who was English. They were the first friends I made in Vienna. They were the people who made me feel like maybe I would be able to stay in town for a little longer. My exit strategy, detailing how to get out of the country with as little embarrassment or fuss as possible, could be postponed for a while.

There was a waitress who knew our names: 'Hello, John!' she would say. 'How was your day today, Wolfie? You've had a haircut, Liam. It looks nice!' It may seem shallow, but it's hard to feel down when there's someone who is smiling and friendly and calling you by your name.

We weren't the only foreign voices there. The staff at 1516 were clearly acutely aware that most of their customers were far away from home, and it's people like them who can help you feel less homesick. A simple 'How are you today?' would make me feel so much more contented. It was a reward for interacting with our new environment.

During that time in Vienna, I lived an almost internet-free life. My only access to email was via the computer in the corner of the staff room or in internet cafés. I did little more than check my emails a couple of times a week and have a quick look at the BBC website when I could. I mainly communicated via telephone boxes and writing letters, which now seems impossibly archaic.

Nowadays, thanks to wifi and smartphones, we have access to the internet in the palm of our hands. But are these comforts that keep us connected useful or damaging? Does seeing what your distant friends are doing exacerbate your fear of missing out or does it make you happier to know that a world so familiar isn't that far away? Can you prevent homesickness happening in the first place?

Dr Miranda van Tilburg has written extensively about homesickness and is the editor of a collection of articles called *Psy-*

chological Aspects of Geographical Moves, all of which focus on homesickness and acculturation stress (the psychological impact of adapting to a new culture).

'It's important to prepare yourself for the eventuality of home-sickness,' she tells me over Skype. 'There will be certain points in the day that cannot be active; they are passive by nature,' she says – times such as eating dinner without a big group around you, or when you're about to go to sleep or have just woken up. 'Those are really, really hard times for people because that's when homesickness will pop up again.'

She tells people to take things from home that are familiar. 'I've known people who would take their own [bedside lamp] or alarm clock because that would be the first thing they would see in the morning.' She also suggests taking a pillow without washing it, so it will smell like home. It's also important to try to have the same routines in your new environment as you had at home, she says.

Can technology help with the potentially problematic initial stages? Perhaps a familiar podcast or downloaded TV programme could be equivalent to the unwashed pillow carrying the smell of home.

'Should we delete our Facebook account or check it in the same way we would do back at home?' I ask van Tilburg. 'In general you would limit how much you use it,' she says, '[do] not check in with your Facebook or Instagram at all times of the day because you will be constantly reminded of home. Do it at one particular time.'

Although it might seem counterintuitive, doing this kind of thing when you are homesick is the worst time to do it. 'It will only increase your feelings of homesickness,' she says, recommending you choose a time of day when you're not usually homesick – maybe during a morning coffee break – and not right before you go to bed. It feels to me like you should check your Facebook when you are happy, rather than when you are sad.

Back on the MIT podcast, Susanna Barry discusses how much contact new students should have with their parents. 'The main guidepost for this is, 'Do I feel like I'm developing my own identity?' 'Do I feel like I am still developing my own friendships, my own way of thinking? Am I able to differentiate my new world and my new identity from my old identity that I had growing up?''

This is what people need to deal with. The moment you say goodbye and the Skype conversation ends, you are left with the black mirror of your tablet reflecting back at you. But while it is easy to criticise technology or say it removes the element of romance, it could actually make a big contribution to your new world away from home.

§ § §

It's about working out the idiosyncrasies. The funny way the tram doors open. Cheating Google Maps by finding a route through the side-streets that shaves minutes off your journey. The difference between ziehen and drücken (don't try to pull a door when there's a big drücken sign in front of you). That's what life's like, living somewhere you're not familiar with. You're constantly trying to pull open doors that are supposed to be pushed, until one day opening the door becomes second nature.

Perhaps homesickness is just an inoculation to make you stronger? Like the Cubs at camp crying themselves to sleep who will, two days later, hate that they have to go home and beg as soon as they get back to be allowed to sleep in a tent in the garden for the night.

As I make my way back into the city from the cottage in the woods, I think maybe there is a cure for homesickness. Maybe you need to have a balance: every time you have a conversation on Skype, you say hello to a neighbour. For every hour you spend on Facebook, you take an hour to check out a flea market or go somewhere different for breakfast. Every time you download one of your favourite podcasts, you try tuning in to a local radio station.

If finding a bar where the staff speak English and there's football on the big screen makes you happy, then do it. The people you love don't want you to be sad, and life is too precious for wallowing. Homesickness is good: it means you're doing it right. But go out and do some things so that next time you speak to the people you miss, you'll do it with a smile on your face and with plenty to tell them. Homesickness can be conquered.

I tell all this to Professor Massey when I speak to her back in the UK. She doesn't like the use of the word 'conquered'. 'You want to think of yourself as a multi-place person [and] incorporate that old place into your identity. 'Conquer it' is a bit brutal. It's as though you're dispensing with one place or another.'

Whatever homesickness is, deep down, all we want is to be with our tribe. But, if we can't, we need to try and create a new one. To find people who know our name in the place we are living, while still having a place we call home, and someone there to tell 'everything's okay'.

This story was first published on 16 June 2015
by Wellcome on mosaicscience.com

Lighting up brain tumours with Project Violet

▉ Alex O'Brien

In 2004, Dr Richard Ellenbogen spent almost 20 hours operating on a 17-year-old girl with a brain tumour. He ended up leaving a big piece of the tumour behind, mistaking it for normal brain tissue. Less than a year after the surgery, the cancer hit back and the young girl died.

The week the girl died, Ellenbogen presented the case at his team's weekly meeting at Seattle Children's Hospital. 'There's got to be a way to take more of the tumour out and leave more of the normal brain intact,' he sighed in frustration. The nagging feeling that he could've taken more tumour out wouldn't leave him alone. Ellenbogen had faced a dilemma: if he had removed more, he would probably have removed more tumour but might also have removed normal brain tissue, with the risk that the girl would have been left severely disabled. Neurosurgeons have to be aggressive and sometimes push themselves to go further and deeper than they feel comfortable going, but they all operate under the adage 'first, do no harm'.

The first recorded cases of cancer show how the Ancient Egyptians used cauterisation (using red-hot instruments to burn off tissue and seal off wounds) to destroy tumours and to treat a variety of infections, diseases and bleeding lesions. Until the mid-18th century, surgery was the only effective option for addressing several conditions. But it was difficult and painful, as shown by the case of Madame Frances d'Arblay, an English novelist living in Paris.

Before operating in 1811, d'Arblay's doctor didn't shield her from the gruelling pain she would encounter during the treatment for her advanced breast cancer – a mastectomy, without anaesthetic. 'You must expect to suffer, I do not want to deceive you—you will suffer—you will suffer very much!' d'Arblay later wrote that 'when the dreadful steel was plunged into the breast—cutting through veins, arteries, flesh, nerves—I needed no injunctions not to restrain my cries. I began a scream that lasted unintermittingly the whole time of the incision... the air felt like a mass of minute but sharp and forked poniards [daggers] that were tearing the edges of the wound.' Yet the operation was a success, and d'Arblay lived for another 29 years.

In 1846, the introduction of ether as an anaesthetic eliminated the pain. Such was its impact that the next hundred years became known as 'the century of the surgeon'. Yet, even in the 21st century, a neurosurgeon removing a tumour still relies largely on just his eyes and touch to guide him.

The differences between cancerous and normal cells are often so minimal that they are extremely hard to tell apart. In the soft, gelatinous mass of the brain, Ellenbogen says tumour cells can feel like 'fruit in Jell-O', meaning that brain tumours are often slightly firmer and have a slightly more leathery texture. Then again, sometimes the tumour can have the same texture as brain tissue. One could distinguish between the two through colour, but even then the difference can be slight. Ellenbogen tells me about one patient whose tumour was barely distinguishable from normal brain matter, except for a hint of yellow in the tumour cells.

There are several imaging technologies that help surgeons see inside the body before they cut, many of which were developed to help diagnose cancer. Ultrasound – bouncing high-frequency sound waves off structures within the body to see what they look like – was pioneered in 1942 by Karl Dussik, a neurologist at the University of Vienna, who attempted to locate brain tumours using essentially the same method bats use to navigate in darkness. On 1 October 1971, the first X-ray computed tomography scan (CT scan) helped to identify a frontal lobe tumour by producing cross-sectional images of the patient's brain. CT scans are good at imaging dense materials like blood and bone, so surgeons will use them if – for example – they are concerned about bleeding in the brain, trauma where bones could be affected or tumours that could involve bones.

One technology commonly used for imaging the inside of the body, especially soft tissues like the brain, is magnetic resonance imaging (MRI). Using a combination of radio waves and a very strong magnetic field, MRI provides information about where the tumour is and how it fits with other important structures in the body. A more 'real-time' version, called functional MRI, has also been used by surgeons to try to map out the safe and unsafe areas to operate on. Since these scans can show roughly which areas of the brain relating to brain function are affected, surgeons are able to give patients a better idea of what they might expect in recovery.

Increasingly, functional MRI is used in combination with intraoperative MRI, a technique developed about 15–20 years ago in which scans are taken at intervals during surgery to help check the progress of the procedure. This helps reduce the risk of incomplete or damaging surgery, says Conor Mallucci, a consultant paediatric neurosurgeon at the Alder Hey Children's Hospital in Liverpool, UK. 'There should be a zero per cent return-to-theatre rate for inadequate surgery,' he adds.

But although all of these technologies have improved things greatly, they are still not precise enough, especially when it comes to brain tumours. We are 'still in the Middle Ages' when it comes to cancer surgery, says oncologist Dr Jim Olson. 'If you look at the rate of individuals that find out that they have bulky cancer left after their cancer surgery, it's staggeringly high – for some cancers like brain cancer, it's as high as 50 per cent. For some very common cancers like breast cancer, it's 30 per cent.'

What excites him is fluorescence imaging – a technology that literally lights up tumours so surgeons can see them. This doesn't just have potential: it's already proving to be effective in providing real-time image guidance to surgeons.

5-ALA, also called Gliolan, is a dye that makes brain tumour cells glow red under UV light. The drug is swallowed by patients 3–4 hours before their surgery, to give it time to accumulate in their tumour cells. It's been approved for use in Europe since September 2007 but has yet to be approved in the USA, where it's undergoing clinical trials; the US Food and Drug Administration has so far refused to approve it, citing that the original trial didn't have overall survival as its primary outcome measure.

'I think it's important for surgeons to have techniques they can use during surgery to identify where they should be removing disease [and], more importantly, to help distinguish areas that they need to leave behind and avoid,' says Dr Colin Watts of the Department of Neurosurgery at the University of Cambridge. Watts is leading a trial looking at whether 5-ALA could also act as a delivery device for a chemotherapy drug (carmustine), which could be implanted into the cavity of the removed tumour to kill leftover tumour cells after surgery.

Fluorescence imaging could allow surgeons to identify tissue that needs to be cut out, such as tumours, and tissue that needs to be avoided, such as blood vessels and nerves. The visual can also help determine whether a tumour can be operated on and, if necessary, what type of follow-up therapy – whether chemotherapy or radiation therapy – would be needed. This added insight allows surgeons and their patients to make more informed decisions about their treatment.

The past few years have seen an explosion of proof-of-concept clinical trials in the field of fluorescent image-guided surgery. However, one method might be even better than 5-ALA, and it's had a lot of attention from both the scientific community and the media. It's called tumour paint.

§§§

Jim Olson remembers being ridiculed. It was 1989 and he was defending his PhD thesis, and the bank of University of Michigan professors asked what his next goal would be. 'If we can bring radioactivity into these tumours for PET scanning, I would love to find a way to bring light into the cancer so that surgeons can see it while they're operating,' Olson told them. The professors chuckled. 'Okay, Buck Rogers,' one of them heckled, 'but what are you really going to do?'

Positron emission tomography (PET) scanning uses radioactive tracers to find cancerous cells. Unlike ultrasound or CT scanning, it detects differences in tissue based not on its structure but on a measure of some metabolic change – the uptake of sugar, for instance. This can help distinguish live tumour from treated tumour, or tumour that's dying. However, surgeons seldom use the technique because it is considered too blunt, unable to give enough precise detail. Moreover, it's not suitable for children because exposing their developing bodies and brains to radiation could lead to other forms of cancer later in life.

A more accurate, radiation-free method that directly targeted cancer cells would be revolutionary, not least for childhood cancer patients. So when Richard Ellenbogen spoke at that Seattle Children's Hospital meeting in 2004, the eyes of Jim Olson, sitting across the table, lit up.

Cancer can occur at any age. Among children under the age of 19, tumours in the brain and central nervous system are the most common cancers. Every year, in the USA alone, approximately 1 in 285 children – nearly 16,000 in total – will be diagnosed with cancer before their 20th birthday, and more than a quarter of them will have been diagnosed with a brain tumour. In most childhood brain tumours, a complete removal improves the overall survival. Olson has diagnosed babies with cancer at one day old, and some even before birth.

The toughest thing about being an oncologist, Olson says, is telling parents – not to mention the child – that a cancer has come back. 'It's one thing to say, 'Your child has cancer and this is the plan, and this is what to expect',' he says. 'It's another thing to begin a conversation that says, 'Your child has recurrent cancer, and it's extraordinarily rare for anybody to survive this...' You're preparing for death and preparing for the miracle at the same time.'

There was no miracle for 11-year-old Violet O'Dell. She had a large, incredibly rare kind of brain tumour. The cancer had woven itself in between the normal nerves of the brainstem – the lowest part of the brain, which connects the brain to the spinal cord and is involved in basic functions such as breathing, heartbeat and blood pressure and some reflex actions like swallowing. It was impossible to remove the cancer without killing her.

The day she was diagnosed was terrifying, Violet's mother Jess tells me, but at the same time it was kind of a relief: there was a sense of 'OK, we know now'. Violet had been falling asleep in class, the beginning of an odd set of behaviour changes. 'For months we had been having problems with her acting odd, belligerent, slurring and stumbling... I would constantly be telling people, I don't know where she went. Whoever this is, this isn't Violet.' After six months of investigation, an MRI scan found the tumour. The diagnosis was given at the Seattle Children's Hospital by Olson.

Olson recalls the conversation he had with Violet when she came to visit his lab. A few days earlier she had seen her mother's driving licence with a heart symbol on it, which indicated that Jess was an organ donor. Violet asked if she could donate her organs too, but Jess wasn't sure because of her cancer.

The then ten-year-old Violet understood that the kind of cancer she had was inoperable, and she knew she was going to die. And after she met Olson's team, who told her about their work, she wanted to know what was being done to research her type of tumour. It was a tough question for a doctor to answer. Olson told her that it was difficult because researchers didn't have material – tumours – to work with. The little girl responded: 'When I die I want you to do an autopsy, take my tumour and put it into mice. So that you can study my cancer and help other kids that get it in the future.'

Violet died a year later at her grandparents' home, beside her family and pets: a puppy, two Labradors and a cat. A couple of weeks after her passing, Olson honoured her by naming a new project 'Project Violet'. He hoped it would revolutionise cancer surgery with a drug from an unusual source.

§§§

Professor Harald Sontheimer, then at the University of Alabama, was never really interested in cancer. If it weren't derived from his favourite cells, glial cells – a type of cell in the nervous system from which most types of brain tumour develop – he probably wouldn't be studying it.

Initially, it was thought that glial cells simply kept everything glued together (the term 'glia' stems from the Greek word meaning 'glue'), but about 30 years ago scientists found out that they do much more. In addition to surrounding neurons and holding them in place, glial cells have three other main functions: supplying nutrients and oxygen to neurons, insulating neurons from one another, and killing pathogens and removing dead neurons. They also help to maintain the blood–brain barrier, which filters substances before they reach the brain, and play a part in the regulation, repair and regrowth of tissue after damage.

There are about 130 different types of brain tumours; the most common type, which is known as a glioma, develops from glial cells. In 1995 Sontheimer and his research team were interested in glial cell-derived tumours. The team had identified that chloride channels – key mechanisms in cells – were somehow involved in tumours invading brain tissue. They were experimenting with substances that would block these chloride channels. And of all of them, the most effective by far was chlorotoxin, a substance isolated from the venom of the deathstalker scorpion.

The deathstalker is one of the most deadly scorpions in the world. Found in arid and hyper-arid regions of North Africa and the Middle East, the Palestine Yellow Scorpion (or Israeli Desert Scorpion, as it is also known) is only 3–4 inches long but can kill creatures a thousand times its size. In humans, the scorpion's sting can cause excruciating pain, convulsions, paralysis and potentially even death (owing to heart and respiratory failure), but most affected humans will just suffer extreme pain in the region of the sting, along with drowsiness, fatigue, splitting headaches and joint pain. These symptoms will sometimes persist for months. They are caused by the scorpion's powerful venom, which is a mixture of neurotoxins – poisons that act on the nervous system – including one known as chlorotoxin.

Sontheimer's team took human brain tumour isolated from a patient and introduced it into the brain of a mouse, then injected synthetic chlorotoxin they had created in the laboratory. What they found was remarkable and unexpected. The synthetic chlorotoxin accumulated in tumours – and only tumours – leaving normal, healthy cells untouched. It also showed a remarkable ability to pass into the brain, unobstructed by the blood–brain barrier. As Sontheimer explained in a later paper: 'They are opening the blood–brain barrier specifically at the point where they're moving along the blood vessels. They're essentially ensheathing the blood vessels, and as they do that, they're gradually breaking down this barrier so that, at the site where they're invading, there is active penetration of this molecule into the brain.'

In 2004 Richard Ellenbogen sent one of his neurosurgery residents, Patrick Gabikian, to Olson's lab at the Fred Hutchinson Cancer Research Center to complete a year of research. There Gabikian would begin looking for compounds that Ellenbogen and Olson could use to illuminate cancerous cells. One day Gabikian came across a compound that Olson saw as a viable option to experiment with: chlorotoxin.

In 2007 Olson, Ellenbogen, Gabikian and their team wrote an article detailing how they isolated chlorotoxin from the death-stalker scorpion venom and attached a fluorescent molecule, producing a substance that would 'light up' tumour cells. Like Sontheimer, Olson's lab uses a synthetic version of the chlorotoxin protein, but the added fluorescent molecule acts like a flashlight – turned on only when it binds to its target.

That target is thought to be a complex containing the protein annexin A2. In normal non-cancerous cells, it lies inside the cell, but in cancerous tissue cells – for reasons that remain unknown – it appears to be turned to the outside of the cell's surface. When chlorotoxin binds, the complex gets moved to the inside of the cell, carrying the flashlight molecule with it. And when that happens, surgeons can direct a laser at the area, emitting a light that can be readily detected by a variety of devices.

By making even the smallest pockets of cancerous cells visible in real-time during surgery, this 'tumour paint' – its official name is BLZ-100 – can help determine the precise location and size of the primary tumour and its satellites (smaller areas of tumour located nearby). This, Olson decided, was a concept worthy of Project Violet. He had the technology. Now what he needed was funding.

§§§

It can take years for government grants to come through, if your application is even successful in the first place. Olson decided a long time ago that applying for grants was just delaying his research. Today, his main financial backing still comes from the families, individuals and family-run foundations that have heard about his research and been interested. He tells me about one father who brought his daughter to the clinic for a flu shot, heard about Olson's work and dropped off a cheque for $100,000.

'Having to say 'no' to Jim is like punching yourself in the face,' says Nicole Pratapas, Philanthropic Gifts Advisor at the Fred Hutchinson Cancer Research Center. She meets with him every Tuesday morning at his office. 'The way that he feels about his patients, and how we do things – how we can raise money [so efficiently] in the amount of time that he spends [doing] it – it means a lot to us.' In the past two years alone, Olson has raised US$6 million dollars through private funding.

That, however, is earmarked for staff costs and the multitude of other research projects he has in the works. When tumour paint started to seem viable, he knew he had a winning idea but had no spare budget to start developing it. What he did have was a short film.

Bringing Light is a three-minute documentary by Bert Klasey, Chris Baron and James Allen Smith. The film was one of 20 finalists at the 2013 Focus Forward Filmmaker Competition and won the Audience Favorite Award. It came about when producer Klasey was looking for ideas and his wife remembered a presentation that had stuck with her since she'd seen it at a conference, eight years previously. It was Olson, talking about what would become tumour paint.

The film presented Olson with an opportunity. He thought he could use it to harness two of the most powerful tools of the modern entrepreneur: goodwill and crowdfunding. As it happens, just a few blocks away from him were the offices of the world's biggest internet retailer, Amazon. He sent them a link to *Bringing Light* and invited the Amazon team to visit his lab, meet the scientists over beer and pizza, and hear more. Olson expected around 20 people would turn up. In fact, more than 70 did (ever the optimist, Olson had catered for 80, 'just in case'). In just a few weeks, 25 of them had formed a volunteer project team that would give up every Monday evening for the next year to design Project Violet's website and put the word out on Facebook and Twitter.

Project Violet is a citizen science project kick-started via an online fundraising campaign, originally based around a drug adoption programme – for a US$100 donation anyone could 'adopt' a potential drug target, which would go toward its research (this has since been removed because, according to Olson, people found it daunting to adopt a drug, wondering if they were smart enough to choose the 'right' one). In less than three years, Olson and his team have raised US$5 million dollars through the Project Violet website and related events, combined with individual donations. Cameron W Brennan, a neurosurgeon at the Memorial Sloan Kettering Cancer Center in New York, views this without cynicism. In fact, he applauds the press coverage: 'Olson has to focus fundraising and to focus research on things that will affect practice.' And Olson has proven before that he is really good at that.

It's taken nearly 25 years, but Olson is on track to make his and Ellenbogen's vision a reality. During preclinical trials with mice and dogs that took place between 2005 and 2011, tumour paint proved to be more than 5,000 times more 'sensitive' than MRI. It could highlight extremely small amounts of cancer cells – as few as 200 cells, whereas MRI has a lower limit of at least half a million.

Unlike 5-ALA, tumour paint can be used in real-time during an operation, and Olson points out that it also crosses the blood–brain barrier, attaches specifically to cancer cells and is internalised. 5-ALA, by contrast, doesn't bind to cancer cells and can only be used for high-grade brain tumours (tumours that are highly malignant and adept at invading nearby brain tissue). Tumour paint has the potential to target lower-grade tumours as well – which is important, says Olson, 'because low-grade gliomas that are incompletely removed often progress to high-grade glioblastomas over the following decade'.

On 25 May 2015, the first clinical trial of tumour paint began at Seattle Children's Hospital, which has the largest paediatric brain tumour centre in the Northwest of America. The trial is run by Blaze Bioscience Inc., and in phase I up to 27 people who have been diagnosed with a brain tumour (from infants to young adults under the age of 30) will have their operations performed using the drug.

'I think lots of other people are trying to find out the difference between brain cancer and normal tissue, but they don't know what to do with the information when they get it,' Olson says. 'They publish their paper and then they move on to the next thing.' He hopes tumour paint will be different.

If it is, it will go some way to living up to Project Violet's name. 'All these buildings and universities get named after wealthy people that donate money,' Olson says. 'I wanted something really beautiful for the world to be named after this little girl.'

This story was first published on 1 September 2015
by Wellcome on mosaicscience.com

The US military plan to supercharge brains

■ Emma Young

In the summer of 2010, Ryan Clark twisted his ankle during a gym class. It was painful, but inconvenient more than anything. He was put on crutches for a week and his ankle healed. Then, six weeks later, the pain returned – only this time, it was a lot worse. Ryan ended up in a wheelchair, unable to bear the agony of walking. Drugs and rehab helped and after six weeks or so he recovered. Then he injured himself again, and a third time, each minor accident triggering pain that became horrendous. 'They were just normal injuries for a nine-year-old,' says Ryan's father, Vince, 'but for him it was huge. As well as the pain, he got tremors. His muscles locked up. He'd go into full body spasms, and just curl up on the floor.'

Ryan was eventually diagnosed with complex regional pain syndrome, a disorder that affects one in a million children his age. Vince Clark, who directs the Psychology Clinical Neuroscience Center at the University of New Mexico in Albuquerque, threw himself into understanding the syndrome and finding ways to help Ryan. Traditional painkillers had provided no relief, so Clark wondered about what he'd been researching in his lab. It's called transcranial direct-current stimulation (tDCS) and it involves applying mild electrical currents to the head.

TDCS belongs to a group of techniques known as 'non-invasive brain stimulation' because they don't involve surgery. It is still experimental, but even in 2010, it was showing promise not only for alleviating pain, but for boosting the brain, improving memory and attention in healthy people. The US Department of Defense (DoD) wondered whether it might benefit military personnel. By the time Ryan became sick, Clark had led DoD-funded studies that explored this question, and produced remarkably good results.

§ § §

The Royal College of Surgeons, London, January 1803. An audience watches in anticipation as the maverick Italian scientist Giovanni Aldini strides into the room. Someone else is on display before them: George Forster, a convicted murderer, who was earlier hanged at Newgate Prison. Using a primitive battery and connecting rods, Aldini applies an electrical current to the corpse. To the spectators' amazement, it grimaces and jerks. In response to rectal stimulation, one of its fists seems to punch the air.

Aldini was fascinated by the effects of electricity on both the body and the mind, Clark tells me. After claiming to have cured a 27-year-old depressed farmer using electrical stimulation, Aldini tried it on patients with 'melancholy madness' at the Sant'Orsola Hospital in Bologna. He had only limited success, in part because the patients were terrified of his apparatus.

Aldini's experiments with electricity were the beginning of a long and storied episode in the history of psychiatry. Electroconvulsive shock therapy, which requires currents strong enough to trigger seizures, was introduced in the late 1930s. But with the rise of effective new drug treatments as well as public criticism in books like Ken Kesey's One Flew Over the Cuckoo's Nest, electrical therapies fell out of favour. 'At some point, our culture became worried about electricity and its effects,' says Clark. 'It was something scary. There was a general anxiety about it, and people weren't willing to look at it in a rational, calm way.'

Clark is animated as he recounts the rise and fall, and subsequent rise, of electrical stimulation of the brain. While the use of electricity on people became frowned upon, neuroscientists still studied the effects on animals – 'A lot of my professors in grad school had played with the effects of electricity in living tissue,' Clark says. In the 1960s, scientists found that tDCS, which involves currents up to a thousand times less powerful than those used in electroconvulsive shock therapy, could affect brain-cell 'excitability' and help with severe depression. But drugs still seemed more promising as psychiatric treatments, so tDCS was abandoned.

Then in the 1980s, electroshock therapy enjoyed a resurgence. It became clear that it could treat some patients with severe depression for whom the drugs did nothing. Around the same time, interest was growing in something called transcranial magnetic stimulation (TMS). A patient undergoing TMS sits very still while a wand held above the skull generates a magnetic field that penetrates their brain. This can relieve depression and also help in rehabilitation after a stroke or head injury.

In 2000, Michael Nitsche and Walter Paulus at the University of Göttingen, Germany, reported that tDCS could alter a person's response to magnetic stimulation. While TMS forces brain cells to fire, tDCS 'primes the pump', as Michael Weisend, a former colleague of Clark, describes it, making it more likely that a brain cell will fire in response to a stimulus.

Neuroscientists' interest in tDCS was reignited by the Göttingen studies. But what really got people talking were the serendipitous findings that tDCS could change the brain functioning not only of patients but also of healthy people, who had been included in the trials only for comparison. This work was hugely influential, Clark says. Researchers began to investigate the potential of tDCS to boost healthy brains. Results showing that it could enhance learning and memory were some of the first to come in. Other teams looked at using tDCS to treat pain. Like many of his colleagues, Clark found it fascinating.

After a postdoctoral role at the National Institute of Mental Health, working in part on TMS, Clark had moved to Albuquerque in a joint appointment with the University of New Mexico and the Mind Research Network (MRN), a non-profit neuroscience research institute. His work focused on brain imaging and schizophrenia. By 2006, he was promoted to Scientific Director at the MRN. Clark was keen to work on tDCS but also needed to get the MRN out of financial difficulties. The institute had over-spent badly. 'We were in a financial black hole,' he says. 'We needed a lot of money fast.'

Around this time, the Defense Advanced Research Projects Agency (DARPA), the part of the DoD responsible for developing new technologies for military use, put out a call for proposals for research in an area they dubbed 'Accelerated Learning'. A general call like this attracts ideas from scientists from across the nation, each hoping that DoD dollars will flood their way. Clark and the MRN got the go-ahead. 'We put a proposal together to use tDCS. And it was funded. And a lot of money came in quickly. A lot of people's jobs were saved.'

It's clear that to Clark, the preservation of jobs by this influx of cash – which ultimately totalled $6 million – helped to justify the use of military funds. He talks positively about the way DARPA does business. 'I do really like their philosophy. They want to promote research that is very cutting-edge and very risky; a 90 per cent failure rate in their portfolio is okay, because the 10 per cent that works will change the world. We got lucky to be in that 10 per cent.'

§ § §

Brian Coffman smiles reassuringly as he leads me into a small room. He's had tDCS done plenty of times, he says, and he's administered it to around 300 people so far. Some report itching, heat and tingling, but nothing serious. Rarely, someone develops a headache.

Coffman, a PhD student who works with Clark, uses adhesive tape to attach the non-stimulating cathode electrode to my left upper arm and the anode, which delivers the current, to the side of my head, up between my ear and my eye. This positioning is designed to maximise the current that is drawn through the target region of my brain. The electrodes are inside sponges that have been soaked in conductive salt water, so a little of the saline drips down my face. They're connected by wires to a 9 volt battery. When Coffman switches on the battery, I feel a tiny spark on my arm. Static discharge, he explains, and apologises.

As Coffman turns the current up to 2 milliamps, the maximum level used in most tDCS studies, I feel a scratchy sensation on my arm, but that's it. Coffman checks that I'm comfortable, then I'm put to work on a computer-based task. The software is called

DARWARS, and it was designed to help familiarise US Army recruits with the types of environments they might encounter in the Middle East. Clark and his team modified it, adding hidden targets to half the 1,200 still scenes. Fairly crude computer-generated images flash up briefly, showing derelict apartment blocks, desert roads, or streets filled with grocers' stands. I have to press buttons on a keyboard to indicate whether there's a threat in the scene or not. Occasionally, it's pretty obvious. Mostly, it isn't. A training period helps the user learn what can be dangerous and what is likely to be benign. When I miss an enemy fighter who's partially concealed, one of my virtual colleagues drops to the dust and I'm verbally admonished: 'Soldier, you missed a threat. You just lost a member of your platoon.'

I didn't feel that the stimulation helped me, though Coffman tells me later that my performance did improve afterwards. This means nothing scientifically – but I can at least attest that while I didn't feel any mentally sharper during or after the tDCS, I didn't experience any negative effects, either.

The MRN team used this software in part of their DARPA-funded research. First, they imaged volunteers' brains to see which regions were active as they learned to spot threats. Then they applied 2 milliamps of direct current for 30 minutes to that crucial region – the inferior frontal cortex. They found that stimulation halved the time it took volunteers to learn. This was a huge surprise, says Clark. 'Most tDCS studies don't achieve a huge effect. A lot are borderline.'

This is one of the criticisms that has been levelled at tDCS: the results aren't always that good. Clark is convinced this is because a lot of the studies haven't involved imaging the brain first, to pinpoint the regions that really need stimulation. 'A lot rely on common knowledge about how the brain is meant to be organised.

I've learned in 33 years of looking at the brain that we still have a lot to learn,' he says. Michael Weisend, who collaborated on the study, agrees – he calls the imaging work 'the secret sauce'.

Despite the impressive results, feedback from colleagues was mixed. And by then, Clark was feeling uncomfortable about several things, not least his benefactors.

§ § §

'It's big. Oh yeah, it's big,' agrees Estella Holmes, an Air Force public affairs representative, who has just driven me in through the gates of the Wright-Patterson Air Force Base in a minivan. Wright-Patt, as it seems to be referred to by anyone who knows the place, is near Dayton, Ohio, and is the largest of all the US Air Force bases, employing some 26,000 people. It is rich in aviation history. In and around this area, Wilbur and Orville Wright conducted pioneering experiments into flight. What they helped to start continues here, at the Air Force Research Laboratory (AFRL).

The AFRL includes the 711th Human Performance Wing, whose mission is to 'advance human performance in air, space and cyberspace'. Wright-Patt is so vast, not even Holmes is quite sure where we're going. We have to ask a passing airman for help. He's dressed in fatigues, even though it's a Monday. On Mondays, Holmes has informed me, it's protocol to wear the blue uniform, unless a grimy task is scheduled. When we get inside, though, everyone seems to be in fatigues. A group of airmen – the term is used for both men and women – are holding an informal meeting at a café in the atrium, while others are walking to their various tasks. Previous Air Force Surgeons General survey the scene from oil paintings hung along one long wall. The atmosphere is quietly busy.

When a young man approaches us, incongruous not only because he's in civilian clothing (a grunge-cool three-piece suit) but because of his long, wavy hair and goatee beard, I'm momentarily thrown. 'When I first met Andy, he looked like he could be active military, while I had a ponytail down to my belt,' Weisend tells me later. 'I like to think I got him on the long-hair path and I'm proud of that!'

Andy McKinley is Weisend's research partner and the military's principal in-house tDCS researcher, leading a lab at the Human Performance Wing. His father was a biomedical engineer in the AFRL. 'I guess I followed in his footsteps,' McKinley says. 'I also liked the fact that my research could lead to the development of technologies that could continue to give us a strategic military advantage and improve national security.' He joined two years after finishing his bachelor's degree and started out investigating the effects of high G-forces on pilots' cognitive performance. After a PhD in biomedical engineering, minoring in neuroscience, he began work on non-invasive (not involving surgery) brain stimulation. 'We began noticing a lot of the medical literature suggesting that cognitive functioning could be enhanced,' he says. 'And particularly in control groups, which were normal, healthy participants. We began thinking: if it could help with those healthy participants, it could potentially be an intervention tool we could use here in the military to help advance cognitive function.'

McKinley has anywhere from six to ten people working on this with him (the number fluctuates according to whether he has summer students or not). And as far as he is aware, his is the only team within the US military, or any other military, investigating non-invasive brain stimulation. Other countries are certainly interested – the UK's Defence Science and Research Laboratory, part of the Ministry of Defence, is paying for research at the

University of Bangor, Wales, on whether tDCS can enhance learning by observation, for example, and for PhD students at the University of Nottingham to conduct studies on enhancing cognition and performance, in part using tDCS.

As a technology, tDCS is unusual in that its effects on healthy people were discovered by accident. So McKinley's research has two prongs. The first is to better understand the basic neuroscience. The second is to develop practical applications.

The day I visit, a tDCS trial is underway in one of McKinley's small labs. An airman sits at a monitor, wired up with electrodes, his jacket slung over the back of his chair. Plane-shaped icons keep entering his airspace. He has to decide whether each incoming plane is a friend or a foe. If it's a foe, he must send a warning. If it flies off, fine. If it doesn't, he must bring it down. The lab is silent, apart from the bleeps as he hits the buttons, and the smash as a software missile destroys an uncooperative plane.

The task obviously involves decision making, but it also has a physical 'motor' component: you must press the buttons in the correct sequence, and you must do this quickly, to get a good score. After a while, this kind of task becomes pretty automatic. 'If you imagine learning to ride a bike or a manual vehicle, your process is very conscious at first because you're thinking about all the steps. But as you do it more often, it becomes more and more unconscious,' McKinley says. 'We wanted to see if we could accelerate that transition with tDCS.'

Brain imaging suggested that the best way to do this would be to stimulate the motor cortex while the volunteer was doing the task. But McKinley and his team added a twist: after the stimulation, they use tDCS in reverse to inhibit the volunteers' prefrontal cortex, which is involved in conscious thinking. The day after the stimulation, the volunteers are brought back for re-testing. 'The results we're getting are fantastic,' McKinley says. People getting a hit of both mid-test and inhibitory stimulation did 250 per cent better in their retests, far outperforming those who had received neither. Used in this way, it seems that tDCS can turbo-boost the time it takes for someone to go from being a novice at a task to being an expert.

In theory, this two-step process might be used to speed all kinds of training, in everything from the piloting of a plane to marksmanship. But for now, image analysis is high on McKinley's list. This is painstaking work that requires a lot of attention. Image analysts spend their whole working day studying surveillance footage for anything of interest.

In other studies, McKinley's team have also used tDCS to supercharge attention, which could help the image analysts too. Volunteers were asked to engage in a rudimentary simulation of air traffic monitoring. Performance at this type of task usually declines over time. 'It's a pretty linear decrement,' McKinley says. But when they stimulated the dorsolateral prefrontal cortex of volunteers' brains, an area they had found to be crucial for attention, they found absolutely no reduction in performance for the entire 40-minute duration of the test. 'That had never been shown before,' he says enthusiastically. 'We've never been able to find anything else that creates that kind of preservation of performance.'

TDCS is not the only brain stimulation tool that he finds interesting. As well as ongoing work into magnetic stimulation, other teams are looking at ultrasound and even laser light, as well as different forms of electrical stimulation, using alternating current, for example. McKinley is about to start looking at ultrasound too, and he's interested in how alternating current can influence brainwaves. But while he says he's agnostic about what type of stimulation might turn out to be best for cognitive enhancement, tDCS has some advantages. For a start, unlike ultrasound or magnetism, electricity is a natural part of brain-cell communication, and it's cheap and portable. He thinks tDCS is the best bet for a wearable brain-stimulating device.

Ultimately, McKinley envisages a wireless cap incorporating electroencephalography (EEG) sensors as well as tDCS electrodes. This two-in-one cap would monitor brain activity and deliver targeted stimulation when necessary – boosting the wearer's attention if it seems to be flagging, for example. The basic technology is already available. And McKinley and Weisend are working to improve and refine it. With help from materials specialists at the AFRL, they have developed EEG-based electrodes that use gel, rather than a wet sponge, and which they say are more comfortable to wear. They also now favour an array of five mini-electrodes within each cathode and anode, to spread the current and reduce the risk of any damage to the skin.

Along with improvements in learning and attention in normal situations, McKinley has found that tDCS can combat the kinds of decline in mental performance normally seen with sleep deprivation. Other researchers have found that, depending on where the current is applied, tDCS can make someone more logical, boost their mathematical ability, improve their physical strength and speed, and even affect their ability to make plans, propensity to take risks and capacity to deceive – the production of lies can be improved or impaired by tDCS, it seems. While much of this work is preliminary, all of these effects may potentially be exploited by any military organisation – though McKinley is at pains to point out that 'soldier mind control' is not what he's about. The biggest barriers to rolling out a tDCS cap for routine use by US military personnel – or anybody else, for that matter – are related not so much to the technology or even the effects it can engender, but to unanswered questions about the fundamental technique.

§ § §

'Let's talk about skulls!'

I'm sitting with Mike Weisend in Max & Erma's, an all-American restaurant about a five-minute drive from his new office at the Wright State Research Institute, which itself is only about ten minutes from the Wright-Patterson Air Force Base. Also at the table are Larry Janning and David McDaniel from Defense Research Associates, a local company that creates technologies 'to support the Warfighter'.

In the car on the way over, Weisend told me about his early, gruesome attempts to get a better idea of what happens to electricity when it's applied to the skull. 'First, I allied with a company that does acoustic damage research on cadaver heads. The idea was we'd get the heads afterwards. It was an incredibly messy, unpleasant business. I couldn't handle it.' But this kind of data is high on his and McKinley's wish list.

No one yet knows what duration of electrical stimulation or what number of stimulations has the biggest impact on performance, or what level of current is optimal. Nor does anyone know whether stimulation might produce permanent change – which might render the two-in-one cap unnecessary, McKinley says, but which may or may not be desirable, depending on the application. There are hints from various studies that even a single session of tDCS might have long-lasting effects. No one knows how long the impacts on attention persisted after the 40-minute cut-off in the air traffic control study, he says.

Another thing nobody knows for sure is where the electricity actually goes when it's applied to various parts of the skull. Certainly, it's a pretty broad, imprecise type of stimulation – a 'shotgun' approach, rather than a 'scalpel', as Weisend describes it. But while there are models that indicate where neuroscientists think the electricity goes in the brain, and so exactly which parts it's affecting, this isn't good enough, says McKinley. You can't put electrodes throughout a living person's head to find out. 'So what we want,' McKinley tells me, 'is a phantom skull.'

Today, Weisend wants to talk to Janning and McDaniel about building this phantom – a model of a human head. The idea is to use a real skull, but with a gelatinous, conductive, brain-mimicking goo inside.

At first, no one's quite sure how to fit the skull with sensors in a way that might produce realistic results, particularly as Weisend wants it to be useful for research with a range of stimulation techniques. Over black-bean burgers and soup, there's talk about multiplex receivers and problems with pulsing signals. Then McDaniel comes up with the idea of inserting a folded fan-type circuit board into the hole at the base of the skull, then opening it up once it's inside. Weisend jumps on the idea. He holds his fists together, the phalanges of his knuckles in contact. 'This is like the brain,' he says. 'You've got fibres running like my fingers.' A fan shape would be a decent mimic for the fibres, he decides. 'I like this idea. I like it a whole lot!'

Both McKinley and Weisend are interested in the basic neuroscience of precisely what tDCS does to the brain, as well as the technology – and the question of safety. This is clearly a big concern when you're talking about zapping the brain with electricity, even if the current is very small. The positive tDCS findings, and the relative cheapness of the kit, has made do-it-yourself tDCS a popular topic for discussion on the internet. You can buy what you need for under $200, and, judging by the online forums, plenty of people are. But Weisend has some major concerns about this. For a start, the electrodes themselves.

'See this?' He rolls up his right sleeve to reveal a small scar on his inner forearm. 'I test all the electrode designs myself before we do it on regular subjects,' he says. 'I don't like to do anything to other people I don't do to myself.' After trying out one particular new electrode, a research assistant wiped his arm and a plug of skin the size of a dime came out. 'It was the consistency of phlegm,' Weisend says. 'I could see the muscle underneath.' The problem was the shape: the electrode was a square, and the current had concentrated at the corners. This was one of many, mostly less unpleasant, results that helped lead McKinley and him to develop the current-spreading five-electrode array.

Nicely packaged consumer tDCS kits, aimed at the public rather than scientists, are already on sale. But Weisend and McKinley – and every other tDCS researcher I've talked to – think it's too early for commercial devices. In fact, they all seem worried. If something goes wrong and someone gets hurt, perhaps by an imperfect electrode design or using the kit for 'too long' – a duration that has yet to be defined – not only will that be regrettable for the individual but tDCS as a concept will be stigmatised, McKinley says.

So far, there seem to be no harmful effects of tDCS, at least, not at the levels or durations of stimulation that are routinely used. Weisend believes there's no such thing as a free lunch, and admits there could be side-effects to tDCS that no one knows about yet. Others are more optimistic. Felipe Fregni, Director of the Laboratory of Neuromodulation at the Spaulding Rehabilitation Hospital in Boston, Massachusetts, says there's no reason to think even long-term use will cause problems, provided that it's at the low levels and durations that are typically used in the lab studies. 'Being a clinician, one thing we are taught at medical school is that treatments that work well have huge side-effects. Then you see something with literally no side-effects, and you think, are we missing something, or not? TDCS is only enhancing what your system is doing. I feel confident that it is pretty safe, based on the mechanisms.'

The absence of side-effects – which most drugs can't boast – is one of the reasons tDCS is so exciting as a clinical tool, says Vince Clark. In many cases, a drug will be more appropriate. But tDCS can relieve pain without making an addict of the user. It can affect the brain without also damaging the liver. As there seem to be no side-effects, tDCS is at least as safe as many drugs that are currently approved for use on kids. Eleven per cent of children in the USA have been diagnosed with attention deficit hyperactivity disorder, and many are on stimulants such as Ritalin. No one knows for sure that there are no very long-term effects of using tDCS – but the same may be said for Ritalin, Clark says.

While tDCS is not approved by the US Food and Drug Administration for any medical use, anecdotal reports lead Clark to believe that its 'off-label' use (when doctors recommend something which they think can help their patient but which isn't officially recognised as a treatment) is growing, particularly for chronic pain and depression. Hospitals are starting to use the technique clinically. In Boston, Fregni and his colleague León Morales-Quezada recently began to use tDCS during rehab on young patients with brain injuries. With one boy, a three-year-old who had suffered severe brain damage after a near-drowning in a swimming pool, they got 'fantastic' results, Morales-Quezada says. After the treatment, the boy had much better control over his movements, and he was able to speak.

There's another 'risk': that the device won't help everyone, and people will say tDCS doesn't work. In fact, people do not respond equally to stimulation, and no one yet knows exactly why. This is just one of the areas that needs more research – which requires money.

§ § §

To Clark, his studies aren't fundamentally about helping to teach a soldier how to spot a threat and deal with it – which, in the real world, might involve identifying and killing an enemy – but about investigating how the brain detects threats. 'A lot of people who've reviewed my work will say that it's good work – but does it have to be about the military? That makes them unhappy. A lot of intellectuals are made uncomfortable by war. Which I am.'

There's something else, which clearly bothers him still. In 2003, Joseph Wilson, a former US diplomat, published a piece in the New York Times arguing that President George W Bush had misled the public about claims of Iraqi purchasing of uranium in Africa, part of the wider furore over the decision to go to war in Iraq. A week later, his wife, Valerie Plame Wilson – a friend of Clark – was outed as a CIA agent. This was retribution, her husband claimed, for his article. 'I'd known Valerie for ten years before this, not knowing she was a CIA agent,' Clark says. 'She was a wonderful patriot, and I was really unhappy that because people were angry at her husband, she lost her career and her ability to do that work... So here were my friends, going through this. And here was I, being pressured to use this technology for weapons development.'

Weapons development? Around the time of the DARPA grant, the focus of the Mind Research Network had begun to shift more and more towards developing tools the military could use, Clark says. 'I'm not allowed to say what was discussed, but I can mention some possibilities,' he says. 'A device that makes enemy troops unconscious, or makes them too confused or upset to fight, might make a weapon. Weapons that alter thoughts or beliefs, or directly affect decision-making or 'reward' pathways in their brain to alter their behaviour, or that keep someone conscious while they are being tortured, might be achieved.' He'd also heard talk of using tDCS to help improve sniper training, which he didn't approve of. 'I had my principles and goals, and they had theirs, and they were in direct conflict.'

In 2009, an error was found in bonus payments to the research assistants on the DARPA project. Clark says that it wasn't that serious, but against the background of his disputes with colleagues over the direction of the institute, it became a big problem. Soon after, he lost his position as principal investigator on the DARPA work.

§§§

After enthusiastic handshakes and promises of further discussions with the men from Defense Research Associates, Weisend yawns, and apologises. He's been in Ohio for only six weeks. It's been a busy period of settling in, getting to know new colleagues and meeting potential collaborators. Also, he and his wife finally got a TV last night, he adds. He couldn't resist staying up to watch old Star Trek episodes. Back inside his office, we sit down and talk about tDCS, his current projects, the Mind Research Network, Vince Clark, the Department of Defense, and the 'colour of money'.

Weisend's cousin David was in the US Special Operations Forces. His sister, Joan, was a career corpsman in the US Navy. She completed numerous tours around the world, including to Iraq and Africa. A shipboard fire on one of her tours resulted in multiple operations on her wrist, neck and shoulder. Between 1997 and 2004, Weisend also worked at the New Mexico Veterans Affairs Hospital, running a magnetoencephalography (MEG) centre, which performed highly detailed scans of patients' brains. He remembers one patient in particular, a woman who'd received a head injury after falling from a moving vehicle during the first Gulf War. As a result, she had epilepsy. MEG scanning of her brain allowed the medical team to perform surgery that stopped the seizures, with the least possible damage to healthy tissue. 'I personally saw the health effects [of military action] on soldiers at the hospital, and my sister, and my cousin,' he says. 'Anything I can do to help those guys and gals, I'll do.'

When Clark lost his position, Weisend was asked to take the lead, and it was he who developed and supervised the second phase of the research. DoD funding forms a big part of his lab income at the Wright State Research Institute, says Weisend – it's for 'exciting, fun' projects he can't talk about. He's well aware that not everyone is comfortable about military-related grants. 'There are people, particularly in university departments, that get worried about the 'colour of money' – Defense money, rather than NIH [National Institutes of Health] money for pure science,' he says. His opinion is that you never know how basic research is going to be used, and if it is used for harm, it's the agency doing the harm that should be open to blame, rather than the researcher who did the original science.

What about the tDCS research on sniper training that Clark had heard about? That belongs to the category of research that has appeared 'in the popular press' but not 'in the lab', Weisend says, though adding that he isn't opposed to it, in theory. 'The bottom line is that Vince and I see the world differently, with respect to the DARPA work and the directions it took,' he says. 'If Vince had conversations about weaponising our results, I was not privy to those conversations. Could the results be weaponised? Undoubtedly. But then again, so could a ballpoint pen. We have always focused on performance enhancement as measured by reducing errors and uncertainty. We never did any experiments on weapons at MRN.'

§ § §

For a long time, it was difficult to get military volunteers for the DARPA-funded studies, Weisend tells me. Unlike civilians, they couldn't be paid for taking part. Then he hit on the idea of ordering a special coin. He passes one over to me. It's weighty and impressive, the size of a medal. On one side is a raised relief of the exterior of a human brain, on the other the full-colour emblems of both the 711th Human Performance Wing and the Air Force Research Laboratory, with 'The Mind Research Network' printed underneath.

Coins like these are really popular within the military, Weisend says. He shows me his collection. There's one from a friend at the Pentagon, another from his cousin, from his time with the 20th Special Operations Squadron of the Air Force, the Green Hornets. 'We couldn't figure out how to get military people in the door,' he says, 'then we came up with these. And they came out of the woodwork to get them.'

While the MRN-led studies involved a mix of military and student volunteers, Andy McKinley recruits his volunteers from the Wright-Patterson Air Force Base. At the moment, tDCS is still experimental, McKinley stresses. It is not yet a routine part of US military training. But some researchers are worried.

Bernhard Sehm, a cognitive neurologist at the Max-Planck Institute for Human Cognitive and Brain Sciences in Leipzig, Germany, has a list of concerns about tDCS and the military. For a start, he says he's far from convinced that lab results would transfer to real-world scenarios, with complex demands – such as combat. Also, 'some researchers have argued that the enhancement of one specific ability might result in a deterioration of another,' he says. 'To use non-invasive brain stimulation in soldiers poses a risk both to the person receiving and to other persons who might be harmed by his actions.' Sehm is also worried about soldiers' autonomy. 'In general, people in the military cannot really decide voluntarily whether to accept a 'treatment' or not,' he says.

As the DoD continues its funding of tDCS research, some researchers in the field have decided to take a firm stand against military-related money. Chris Chambers, a psychologist at Cardiff University, in Wales, conducts research into magnetic brain stimulation. When he was approached by representatives from QinetiQ, a British defence technology firm, who told him that funding might be available for joint collaborations, he says he rejected their overtures, on a point of principle.

This isn't necessarily an easy decision. Pharmaceutical companies aren't interested in paying for the research, because not only is tDCS not a drug but in some cases it could be in direct competition with a drug, and may even have big advantages. 'It doesn't circulate through the body, so it won't affect other organs that most drugs can damage,' Clark says. 'It's not addictive. If there's any problem, you can turn it off in seconds. It's also cheap.' These benefits, unfortunately, restrict researchers' options to public funding bodies (who haven't exactly thrown money at tDCS), private defence-related companies, or the military.

In the past, DoD funding has produced innovations that have had a huge impact on civilian life – think of the Global Positioning System of satellites or even noise-cancelling headphones. Andy McKinley hopes a safe, effective form of tDCS will join that list. While the DoD doesn't have enough in-house specialists to do the research, it does have cash.

Clark still acts as a research supervisor at the MRN, but works mostly at the university. He is currently gathering 'whatever little pieces of money I can find' to pursue medical-related research: to investigate whether tDCS can cut drinking in alcoholics, reduce hallucinations in people with schizophrenia, and calm impulsive behaviour associated with fetal alcohol spectrum disorder. While this research is relatively cheap, funding is still a problem. Given the recent rapid rise in tDCS research published in academic journals, Clark hopes the NIH will soon start taking tDCS research seriously, and pay for large-scale, controlled studies.

Among the promising leads are further findings that tDCS also seems to work well with types of pain that don't respond well to conventional painkillers, like chronic pain, and pain from damaged nerves. In these cases, the target is usually the motor cortex, and the idea is to reduce pain signals. Which brings me back to Ryan, one of the biggest motivations for Clark's research. Did Clark eventually try it on his son? When Ryan first got sick, 'none of the doctors here had heard of tDCS,' he tells me, 'and without medical help, I decided I wasn't going to do it'. He also came across a low-tech approach: an 'orthotic', similar to the mouthguards people used to stop night-time teethgrinding. To Clark's surprise, this relieves Ryan's pain and eases his movement. But Clark says he'd be happy for Ryan to try tDCS. If the mouthguard stopped working and he could find a clinician who would work with the technique, 'I don't think it would be any problem'.

Clark raves about its potential to aid sick people, like his son, and healthy people alike. But he says he's clear now about his position on what funds to accept and what research to do. 'I want to see tDCS used to help,' he says, 'not to harm.'

This story was first published on 3 June 2014
by Wellcome on mosaicscience.com

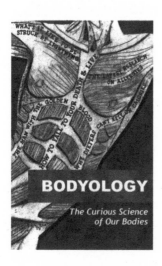

BODYOLOGY
The Curious Science
of Our Bodies

ISBN: 9780995497863
Paperback
Pages: 272
RRP: £8.99

Also
9780995497870 Ebook
9780995497887 Audiobook

Acknowledgements

The stories in this book were originally published by Wellcome on mosaicscience.com. Canbury Press acknowledges the people who worked on these stories for Mosaic, including:

John Walsh, Louisa Saunders, Liana Aghajanian, Rob Reddick, Charlie Hall, Sam Wong, Mun-Keat Looi, Lowri Daniels, Tom Freeman, Peta Bell, Geoff Watts, Jim Giles, Linda Geddes, Olivia Solon, Chrissie Giles, Francine Almash, Jo Marchant, Andrea Volpe, Srinath Perur, Mike Herd, Giles Newton, Audrey Quinn, Madeleine Penny, Will Storr, Lucy Maddox, Jennifer Trent Staves, Shayla Love, Michael Regnier, Gaia Vince, Michael Regnier, Laura Dawes, Katherine Mast, John Osborne, Kirsty Strawbridge, Alex O'Brien, Emma Young, Cameron Bird, and Frieda Klotz.